ARE WE BORN RACIST?

ARE WE BORN RACIST?

New Insights from Neuroscience and Positive Psychology

Edited by Jason Marsh, Rodolfo Mendoza-Denton, and Jeremy Adam Smith

Beacon Press
Boston

Beacon Press
25 Beacon Street
Boston, Massachusetts 02108-2892
www.beacon.org

Beacon Press books
are published under the auspices of
the Unitarian Universalist Association of Congregations.

Printed in the United States of America

13 12 11 10 8 7 6 5 4 3 2 1

Text design and composition by Wilsted & Taylor Publishing Services

Library of Congress Cataloging-in-Publication Data
Are we born racist? : new insights from neuroscience and positive psychology / edited by Jason Marsh, Rodolfo Mendoza-Denton, and Jeremy Adam Smith.
p. cm.
ISBN 978-0-8070-1157-7 (pbk. : alk. paper) 1. Prejudices. 2. Racism—Psychological aspects. I. Marsh, Jason. II. Mendoza-Denton, Rodolfo. III. Smith, Jeremy Adam, 1970–
BF575.P9A74 2010
305.8001'9—dc22 2009039744

All other chapters were originally published in slightly different form in *Greater Good*, the magazine of the University of California at Berkeley's Greater Good Science Center.

Editors' Note

Are We Born Racist? *is not an academic book, but most of the essays that appear in this volume are based on peer-reviewed scientific studies published within the past ten years. While the editors have elected not to provide endnotes, interested readers can seek out the selected studies and resources (many of which are freely available on the Web) listed at the end of the chapters written by scientists in order to understand the discoveries and methodologies that inform these contributions.*

CONTENTS

PART I THE NEW PSYCHOLOGY OF RACISM

PART II OVERCOMING PREJUDICE

PART III **STRENGTHENING OUR MULTIRACIAL SOCIETY**

PART I

THE NEW PSYCHOLOGY OF RACISM

Introduction

The Editors

In 2007 economists Joseph Price and Justin Wolfers found that National Basketball Association referees are more likely to call fouls on players of a different race from themselves. Price and Wolfers's findings were roundly dismissed across the league. Rod Thorn, the league's former vice president of basketball operations, flatly stated, "I don't believe it." One veteran player, Mike James, remarked, "If that's going on, it's something that needs to be dealt with . . . but I've never seen it."

James makes a good point: such statistical evidence is often very difficult to perceive with the naked eye; as a result, there is a tendency to dismiss such data as coincidence. Yet the point of this finding is that it represents not single games, individual referees, or individual players, but rather stable trends that summarize the behavior of many individuals over a large number of foul calls—in this case, over a half million through the course of thirteen professional seasons. In 2007 a similar analysis of Major League Baseball data reached a strikingly similar conclusion: that umpires are more likely to call a given pitch a strike when the pitcher is of their same race. Here, the database consisted of more than a million pitches over the course of two seasons. It becomes difficult to make a case for coincidence.

These studies are about a lot more than foul calls and strikeouts. They're about the nature of racism today: subtle, pervasive, persistent. And they force us to consider some uncomfortable questions. If umpires and referees, who are professionally trained to avoid bias, are still subject to racism, what hope is there for the rest of us? Is racism more deeply rooted in the human psyche than we'd like to admit? Are we, in fact, born racist?

These are the questions that animate this book, and they're the questions that drive the latest research into the sources and consequences of prejudice. *Are We Born Racist?* collects essays from *Greater Good* (www.greatergoodmag.org), a magazine published by the Greater Good Science Center at the University of California, Berkeley, on what new scientific discoveries are revealing about the nature of prejudice—and about how we can overcome it. We devote special attention to racial prejudice, but most of the research covered in this anthology also applies to prejudices based on other characteristics that include age, gender, religion, and sexual orientation.

Are We Born Racist? brings together writers and scientists to help us make sense of an emerging science of racial prejudice. This research includes findings from neuroscience that reveal the brain mechanisms behind our responses to people of different races, as well as findings from positive psychology that highlight our equally powerful impulses for empathy, compassion, and fairness. Although these strands of research may seem incompatible at first, a closer look reveals a complex yet fascinating interplay between prejudice and egalitarianism that, ultimately, gives us reason to keep up the fight against racism. As Susan Fiske, Kareem Johnson, and David Amodio suggest in their contributions, neuroscience has discovered racial prejudice rooted in brain areas that emerged early in primate evolution and that still govern our instincts today. But research has also revealed how the more recently evolved neocortex functions to regulate our automatic impulses and helps us achieve our goals: eating only one slice of chocolate cake, planning the best way to ask for a raise, or being fair to people who are different from us.

While neuroscience provides us with the tools for understanding our minds, positive psychology reveals the human capacities that can be put to use in building an authentically multicultural society. In combining and integrating the latest research from these subfields, *Are We Born Racist?* offers a new angle for scientific discussions around inequality and discrimination. Science has always held a special role in shaping our attitudes and behavior toward those of other races. Unfortunately, this role has not always been positive.

From measurements of head circumference to eugenics to discussions around IQ distributions, science is too often used to contribute and support discrimination. The time is now ripe to marshal the latest scientific research in the fight against racism.

Of course, this fight must remain multifaceted, because racism does not begin and end in the brain. Instances of structural racism are all around us. Predominantly black, low-income neighborhoods are targeted by cigarette companies and the liquor industry for billboards and stores. Minority homeowners are overrepresented in the subprime, high-interest loan market. Huge racial disparities remain in the quality of students' education, and in their levels of achievement. These systemic inequities are clearly more wide-ranging than individual instances of racial prejudice, but are nevertheless maintained and perpetuated by the decisions and actions of individuals. The essays in this volume center on the psychology behind racial prejudice—the individual acts and interpersonal interactions that, like so many called fouls and strikes, imperceptibly yet powerfully snowball into unmistakable patterns of racism.

Like referees and NBA officials, people usually resist the notion that they might be perpetrators of prejudice. Their resistance often rests on the fundamental assumption that prejudice and racism are all-or-none qualities, where one is either racist or not. Yet this assumption leaves no room for the possibility that one might sincerely hold egalitarian goals and *simultaneously* be at risk for perpetrating racism. Recall the admission Jesse Jackson made in a 1993 interview: "There is nothing more painful to me at this stage in my life than to walk down the street and hear footsteps and start to think about robbery and then look around and see somebody white and feel relieved." With this quote, Jackson recognizes that prejudiced thinking can sometimes occur outside of our control—a phenomenon that cuts across racial categories, and one that writer Rona Fernandez tackles in her contribution to this volume.

Indeed, this all-or-none assumption about the nature of racism makes the possibility of automatic or unconscious prejudice extremely threatening—particularly if egalitarianism is one of our core values. And research shows that we are awfully good at explaining away information that threatens our most cherished beliefs and values (including those about ourselves). So, paradoxically, people who most cherish egalitarianism might be at greatest risk for denying the persistence of racism, as Rodolfo Mendoza-Denton discusses in "People Understand Each Other by Talking."

Yet one of the worst ways to advance in the battle against racism is to deny the possibility of its continued existence: we cannot address and control behavior that we say doesn't exist. It is possible to overcome subtle prejudice, but first we have to recognize that the problem remains.

In the first part of this anthology, "The New Psychology of Racism," scientists highlight why and how our brains form prejudices and how racism hurts our health—and they suggest what steps we can take to mitigate racist instincts. In the second part, "Overcoming Prejudice," journalists and researchers apply the newest tools of psychology to overcome bias at school and work, and in our interpersonal interactions. Then in the third and final part, "Strengthening Our Multiracial Society," we ask writers and social scientists to reflect upon the implications of the research for daily life, and to take the discussion to the next level: How might we grow to accept the diversity we see around us and also inside of us? How can we go beyond mere tolerance, to truly embrace racial and cultural differences?

That may sound like a utopian goal, but we believe that the new neuroscience of racism suggests new ways to live. These discoveries should provoke all of us to think about the next stage of our development—as individuals and as a society.

Are We Born Racist?

Susan T. Fiske

How prejudiced are you?

Most people think they're less biased than average. But just as we can't all be better than average, we can't all be less prejudiced than average. Although the success of Barack Obama's presidential campaign suggests an America that is moving past traditional racial divisions and prejudices, it's probably safe to assume that all of us harbor more biases than we think.

Science suggests that most of us don't even know the half of it. A twenty-year eruption of research from the field of social neuroscience reveals exactly how automatically and unconsciously prejudice operates. As members of a society with egalitarian ideals, most Americans have good intentions. But new research suggests our brains and our impulses all too often betray us. That's the bad news.

But here's the good news: more recent research shows that our prejudices are not inevitable; they are actually quite malleable, shaped by an ever-changing mix of cultural beliefs and social circumstances. While we may be hardwired to harbor prejudices against those who seem different or unfamiliar to us, it's possible to override our worst impulses and reduce these prejudices. Doing so requires more than just individual good intentions; it requires broad social efforts to challenge stereotypes and get people to work together across group lines. But a vital first step is learning about the biological and psychological roots of prejudice.

Modern prejudice

Here's the first thing to understand: modern prejudice is not your grandparents' prejudice.

Old-fashioned prejudices were known quantities because people would mostly say what they thought. Blacks were lazy; Jews were sly; women were either dumb or bitchy. Modern equivalents continue, of course—look at current portrayals of Mexican immigrants as criminals (when, in fact, crime rates in Latino neighborhoods are lower than those of other ethnic groups at comparable socioeconomic levels). Most estimates suggest such blatant and wrongheaded bigotry persists among only 10 percent of citizens in modern democracies. Blatant bias does spawn hate crimes, but these are fortunately rare (though not rare enough). At the very least, we can identify the barefaced bigots.

Our own prejudice—and our children's and grandchildren's prejudice, if we don't address it—often takes a subtler, unexamined form. Neuroscience has shown that people can identify another person's apparent race, gender, and age in a matter of milliseconds. In this blink of an eye, a complex network of stereotypes, emotional prejudices, and behavioral impulses activates. These knee-jerk reactions do not require conscious bigotry, though they are worsened by it.

In my own lab, for example, we dug up dozens of images of societal groups that were identifiable in an instant: people with disabilities, older people, homeless people, drug addicts, rich businessmen, and American Olympic athletes. We asked research participants to tell us what emotions these images evoked in them; as we predicted, they reported feeling pity (toward the disabled and elderly), disgust (the homeless and drug addicts), envy (businessmen), and pride (athletes).

We then slid other participants into a functional magnetic resonance imaging (fMRI) scanner to observe their brain activity as they looked at these evocative photos. Within a moment of seeing the photograph of an apparently homeless man, for instance, people's brains set off a sequence of reactions characteristic of disgust and avoidance.

The activated areas included the *insula*, which is reliably associated with feelings of disgust toward objects such as garbage and human waste. Notably, the homeless people's photographs failed to stimulate areas of the brain that usually activate whenever people think about other people, or themselves. Toward the homeless (and drug addicts), these areas simply failed to light up, as if people had stumbled on a pile of trash.

We were surprised—not by the clear sign of disgust, but by how easy it was to achieve. These were photographs, after all, not smelly, noisy, intrusive people. Yet we saw how readily physical characteristics could evoke strong, immediate, and deep-seated emotional reactions.

Results like these have obvious implications for racial prejudice, which is often elicited by similarly superficial characteristics. Indeed, a great deal of recent research has shown how our knee-jerk biases are directed toward members of other races.

Research by New York University psychologist Elizabeth Phelps and her colleagues has found that even dull yearbook photographs can trigger a strong neural response. When white men in their study briefly saw pictures of unfamiliar black male faces, their brain activity spiked in a region known as the amygdala, which is involved in feelings of vigilance generally, and in the fear response specifically; the amygdala lights up when we encounter people or events we judge threatening. Several other labs, including my own, have uncovered a similar link between amygdala activity and white people's perceptions of black faces.

Other research has uncovered more subtle forms of racial bias. In one study neurosurgeon Alexandra Golby and her colleagues showed participants images of white and black faces. When white participants saw white faces, their brains showed more activity in a region that specializes in facial recognition than when they saw black faces; the same went for black participants when they saw black faces. For some reason, those other-race faces didn't register as human faces in the same way that same-race faces did. Later, all participants saw a series of white and black faces, some of which were new faces and some of which were faces they'd already seen during the brain scans. Sure enough, both white and black participants proved better able to remember people of their own race.

Work by Stanford psychologist Jennifer Eberhardt and her colleagues suggests that these rapid, unconscious facial perceptions can have deadly consequences. The researchers had participants analyze photos of African American men convicted of murder, rating how "stereotypically black" the men's facial features appeared. Some of the men had been sentenced to the death penalty; some had been given less-severe sentences, though the participants didn't know which men were which. Even after controlling for relevant variables such as the severity of the murder and the defendant's facial attractiveness, socioeconomic status, and prior convictions, it turned out that black men were more than twice as likely to be carrying a death sentence if they had facial features that were judged by study participants as more "stereotypically black." (However, this discrepancy only existed if the murder victim was white. The defendants were no more likely to get the death penalty if their victim was also black.)

Meanwhile, in studies mimicking how the police deal with criminal suspects, University of Chicago psychologist Joshua Correll and colleagues have shown that police officers, community members, and students playing a video game are faster to "shoot" an armed black man than an armed white man, and they're faster to avoid shooting an unarmed white man than an unarmed black man. (For more on this line of research, see Alex Dixon's "Policing Bias" later in this anthology.) Cultural stereotypes and emotional prejudices register on the brain as quickly as a fifth of a second—enough time to determine whether a suspect lives or dies.

Us versus not-us

Years before these neuroscience findings, social psychologists had documented the instant (and unfortunate) associations people make toward "out-groups"—those groups they don't consider to be their own. Whether they differ by age, ethnicity, religion, or political party, people favor their own groups over others, and they do so automatically. We have always had codes: PLU (people like us), NOKD (not our kind, dear), the 'hood, the Man. Every culture names the "us" and the "not-us." It appears to be human nature, and many studies have shown how easy it is to provoke this kind of psychological distinction between our "in-groups" and "out-groups."

In one of the most famous of these studies, pioneering social psychologist Henri Tajfel showed teenage boys paintings by Klee and Kandinsky and asked them which artist they preferred. Tajfel then gave the boys the chance to distribute money to others who preferred the same artist, or to those who liked the other artist. The Klee boys were significantly more likely to give money to other Klee fans; Kandinsky boys were significantly more likely to share with other Kandinsky-ites. They proved decidedly loyal to their groups, even though they'd become affiliated with this group just minutes earlier, knew nothing else about their fellow group members, and ostensibly had nothing to gain from their group membership.

Similar studies have shown that people demonstrate strong preferences toward those wearing the soccer jersey of a team they like, people who share their birthday, and people who subtly resemble themselves, not to mention those of their own race or ethnicity. Conditioned by millennia of tribal warfare and fierce competition for limited resources, we are always looking for cues to help us make snap judgments about others.

Unfortunately, as we gravitate toward the familiar and the similar, all too often we rely on physical characteristics to determine whether someone is in our in-group or out-group. In that light it's not hard to understand why so much prejudice is directed at people based on their race.

What's more, we all have to contend with our culture's influential role in shaping prejudice. Years, even generations, of explicit and implicit cultural messages—gleaned from parents, the media, firsthand experiences, and countless other sources—link particular physical appearances with a host of traits, positive or negative. The roots of these messages can stretch back centuries, as is the case with racism toward people of African descent in the United States and its origins in the age of slavery. Such messages are absorbed, accepted, and perpetuated, often unconsciously, by our culture's members and institutions. That's how prejudices become so widespread and automatic.

A fight we can win

People have a tendency to think that biology is destiny. But just because we can correlate impulses in the brain with certain prejudices does not mean we are hardwired to hate drug addicts and homeless people, or that members of different races are destined to fear and mistrust one another.

In the neuroscience studies looking at race, for instance, amygdala (vigilance-related) reactions vary by individual, corresponding to other signs of prejudice. People who exhibit more prejudiced attitudes or behaviors, for example, show more amygdala response. And the alarms in whites' amygdalae do not go off to famous black faces. Likewise, their brains grow accustomed to new black faces after repeated exposure.

Ohio State researcher William Cunningham has even found that, among whites, black faces trigger more amygdala activity only when these faces were seen for a length of time (thirty milliseconds) so short that it amounts to subconscious exposure. When whites had the chance to see black faces for a bit longer (525 milliseconds) and process them consciously, their amygdala activity wasn't unusually high; instead, they showed increased activity in brain areas associated with inhibition and self-control. It was as if, in less than a second, their brains were reining in unwanted prejudices.

The most important lessons of this whole wave of research point to the complexity of the interactions between biology and environment.

Take the amygdala race results. When researchers just slightly change the social context in which people view photos of other races, we've seen changes in the ways their brains react to these faces.

In my own lab, for instance, we showed white study participants a series of photos, some of white faces and some of black ones. We gave them two seconds to answer one of three questions about the people in these photos: whether they were over twenty-one, whether they had a gray dot on their face, or whether they liked a certain vegetable. When participants had to decide if the people in the photos were over twenty-one, we saw a spike in their amygdala activity, similar to what had been found in the studies I mentioned earlier. But when they looked at these faces to judge what kind of vegetable the person would like, or when they were looking for a gray dot, their amygdala activity was the same as when they saw white faces.

In other words, when our study participants had to place others into a social category—even if it was by age, not race—they saw black faces differently than white faces. But the gray dot exercise showed it was possible for whites to look at black faces without getting this effect. More important for everyday interactions, when participants were prompted to judge these people as individuals—individuals with their own unique tastes and preferences—they reacted no differently to black faces than they did to white ones.

Similarly, my lab's latest brain scans indicate that people stop dehumanizing homeless people and drug addicts when they're made to guess what these people would like to eat, as if the study participants were running a soup kitchen.

What this research suggests is that the environment can interact with human nature for good or ill; social conditions can reduce prejudice, just as they can foster or exacerbate it.

Both science and history suggest that people will nurture and act on their prejudices in the worst ways when these people are put under stress, pressured by peers, or receive approval from authority figures to do so. We see this in hate crimes directed at those who are homeless, people perceived to be gay or lesbian, and people of all ethnicities; my former students Lasana Harris, Amy Cuddy, and I have argued that these processes lie at the root of prisoner abuse in settings such as Abu Ghraib.

Fortunately, research has also indicated which kinds of social conditions can reduce prejudice. For instance, a long line of my previous research indicates that putting people on the same team helps to overcome prejudices over time. In one study my former student Steve Neuberg and I found that study participants had negative feelings toward a schizophrenic patient recently discharged from a mental institution—unless they were told they'd have to work with him for a chance to win a significant monetary prize. Then they noticed and judged him more by his own unique, individual traits, not by the traits associated with his stigmatized group.

Our results echo the famous "Robbers Cave" experiment led by Muzafer Sherif, a founder of social psychology. Sherif brought two groups of boys to separate parts of a campground and encouraged each group to bond as a team, not telling them about the other group at first. As both groups became aware of the other one, a fierce rivalry developed between them. Yet Sherif and his colleagues soon posed a series of challenges to the groups that neither could solve without the help of the other. As they started to work together, their old tensions dissipated and they bonded across group lines.

These findings are part of a long line of research supporting what's known as the contact hypothesis, which states that under the right conditions, contact between members of different groups can reduce conflicts and prejudices. Decades of school desegregation research support this idea, as documented by University of California at Santa Cruz professor emeritus Thomas Pettigrew and University of Massachusetts at Amherst psychologist Linda Tropp.

Pettigrew and Tropp have found that school integration can in fact reduce prejudice among students from different groups, but simply placing these students together isn't enough to get them to see each other as individuals and shed their prejudices. We must also try to help them share common goals, on which they must cooperate to succeed; ensure that they're treated as equals and have positive, noncompetitive interactions with one another; and show that their cross-group relationship has the support of authority figures. The more of these factors in place, the more likely people are to overcome their biases. This has proven to be true not only in schools but in a variety of other social institutions, from the military to public housing projects. Our biases are not so hardwired after all, given the right social engineering.

As a society, we engage in social engineering all the time, mostly by accident and without intending or even anticipating all the consequences. For example, we admit athletes and the children of alumni to colleges under unexamined kinds of affirmative action that dwarf racial affirmative action. Colleges even practice affirmative action for high school boys, whose grades and test scores on average are lower than girls'. Affirmative action by race is a more-examined form of social engineering, but it is only one among many.

Because of other forms of social engineering—the kind perpetrated by biased real estate agents and job interviewers—today we remain racially segregated in our neighborhoods and workplaces. This holds true even after accounting for social class. As a result, people are deprived of daily interactions with others who might seem superficially different from themselves, but who in fact share the same values, hopes, and fears.

When we allow our society to remain so segregated, as documented by my husband, sociologist Doug Massey, then all else follows: the less fortunate are exposed to violence and disorder, receive an inferior education, have fewer local job opportunities, and lack constructive role models. Once we understand how automatically people fear difference, dehumanize the less fortunate, and demonize the Other (or "not-us"), we can better appreciate how these forms of segregation can perpetuate themselves, and why we must fight against them. The science of human prejudice suggests that, if we're informed and persistent, this is a fight we can win.

Further Reading

Fiske, S. T., H. Bergsieker, A. M. Russell, and L. Williams. 2009. Images of Black Americans: Then, "them" and now, "Obama!" *Du Bois Review: Social Science Research on Race* 6:83–101.

Macrae, C. N., and G. V. Bodenhausen. 2000. Social cognition: Thinking categorically about others. *Annual Review of Psychology* 51:93–120.

Pettigrew, T. F., and L. R. Tropp. 2006. A meta-analytic test of intergroup contact theory. *Journal of Personality and Social Psychology* 90:751–83.

Phelps, E. A., K. J. O'Connor, W. A. Cunningham, E. S. Funayama, J. C. Gatenby, J. C. Gore, and M. R. Banaji. 2000. Performance on indirect measures of race evaluation predicts amygdala activation. *Journal of Cognitive Neuroscience* 12:729–38.

Wheeler, M. E., and S. T. Fiske. 2005. Controlling racial prejudice: Social cognitive goals affect amygdala and stereotype activation. *Psychological Science* 16:56–63.

Prejudice versus Positive Thinking

Kareem Johnson

Revelations in social neuroscience have shown that racial categories are the first thing that we notice about a new person, before their age, gender, or other social characteristics. But, as Susan Fiske describes in the first essay of this volume, the research also suggests that the very act of focusing on someone's racial category can distort our perception of the faces we see. This focus even affects the way our brains process faces: when we see the face of someone of a different race, our brains don't respond as they do when we see the face of someone of our own race. It's almost as if our brains are wired to dehumanize the faces of people who belong to a different race.

It's also the case that our own faces can betray prejudice, even when we don't want them to. Our facial expressions and gestures reflect what's going on in our brains; in turn, those expressions and gestures affect what happens in the minds of other people.

As an African American and as a human being, I find many of the results from these studies dismaying. However, as a social psychologist I know that there is more to the story. I have discovered that although the influence of racial categories and stereotypes is formidable, it is not impossible to overcome. Prejudice isn't all the body can reveal. In our gestures and expressions, researchers have also discovered a powerful desire to be truly egalitarian. In my own studies I've found that prejudiced impulses and perceptions can be dispelled by something as simple as a smile.

Betrayed by body and brain

In the wake of the successes of the civil rights movement, egalitarianism and racial tolerance have assumed a larger role in our moral ethics. The term *racist* has become an epithet in most parts of American society. It is rare and unusual for someone to openly report bigoted feelings. While this change is welcome, it has caused a slight problem for psychologists who study how stereotypes and prejudice influence behavior. In the past decade a great deal of research has been dedicated to measuring thoughts and feelings about race that may lie beneath the surface. Psychological tests that measure these more *implicit* racial attitudes often examine how easy it is to associate people of a different race with positive or negative words or objects.

As far back as the 1950s, researchers were examining differences between what we say during interracial encounters and what we may actually feel during those encounters. Studies in psychophysiology examine how our bodies respond to stress and emotions. A lie detector test, or polygraph, is a common example. When someone makes an attempt at deception, they will often feel stress from the fear of getting caught. Of course, they aren't going to tell you that. But their bodies will give it away: their hands may start to sweat more, their hearts may beat faster, their breathing may speed up—perhaps not enough for you to notice with your naked eye, but the machine can tell the difference. Early studies using psychophysiology to examine interracial attitudes found that during real or imagined interracial interactions, people often showed bodily signs of stress inconsistent with their explicitly stated racial attitudes.

Those studies began over fifty years ago, but recent research shows much the same thing. For example, psychologist Eric Vanman, currently at the University of Queensland, used facial electromyogram (EMG) sensors to examine subtle facial expressions while study participants imagined an interracial interaction. He found that white students would consistently rate potential black partners as favorably or more favorably than potential white partners. But what they explicitly reported was often in direct conflict with their facial expressions.

When imagining interacting with a white partner, the students were likely to smile a little—but when imagining interacting with a black partner, they were more likely to furrow their brows, an expression linked to disapproval.

In another study Yale University psychologist John Dovidio and his colleagues examined how people's racial attitudes predicted verbal and nonverbal behavior during same-race and cross-race interactions. When he videotaped a series of interactions and later had his research team rate how friendly the interaction was, he found that the racial attitudes people explicitly reported on a survey predicted how verbally friendly they were during the interaction. Caucasians who held more egalitarian beliefs had friendlier conversations with African American interaction partners.

Then Dovidio reexamined videotapes of the interactions without any sound, so he could rate how friendly the interactions were in terms of nonverbal body language. Regardless of their explicit racial attitudes, implicit racial biases predicted nonverbal behaviors. People who wished to present themselves as nonprejudiced carefully chose their words to make the interaction friendly, but their body language revealed underlying feelings of stress and discomfort.

Smiling against prejudice

Does this mean that we are doomed to be racist? Although at first glance this research appears to paint a bleak picture, social psychologists are also discovering that under the right circumstances, we can avoid the influence of automatic and unconscious stereotypes.

Recent work in my own lab has shown that the proper social context can alter implicit racial biases. Keith Payne at the University of North Carolina at Chapel Hill has shown that people are faster to recognize negative objects like guns after seeing black male faces because of the strong associations between black men, aggression, and crime. However, using a modified version of Payne's gun-identification task, my lab has recently shown that this association can be altered by the facial expressions shown on the faces. We found that the tendency to see guns disappeared when the black faces shown were smiling. It seems that the positive associations with expressions like smiling were enough to weaken immediate associations between black men and guns.

There's another way smiles can help reduce implicit biases. In a clever study by Tiffany Ito and colleagues at the University of Colorado, Boulder, research participants were unknowingly made to smile while looking at pictures of unfamiliar faces of Caucasians and African Americans. How do you make someone smile unknowingly? Easy. While the participants were viewing the faces, they were instructed to hold a pencil in their teeth and not let it touch their lips. Go ahead and try it yourself. Although it's not a real smile, the face that you make while holding the pencil in your teeth feels like a smile. Past research has found that this kind of facial feedback, posing a smile, can be enough to produce mild positive emotions and can lead to more favorable evaluations of other people.

Ito and colleagues found that participants who "smiled" when they viewed a set of African American faces showed significantly less racial bias on a later test of implicit attitudes. Apparently, the act of smiling while viewing the faces allowed for more positive associations with black faces, resulting in more positive implicit attitudes.

Recently, Sophie Lebrecht of Brown University and Jim Tanaka of the University of Victoria, British Columbia, developed a training method that can weaken implicit racial biases and improve recognition of other-race faces. At the beginning of the study, Lebrecht and Tanaka measured Caucasian participants' implicit racial attitudes and ability to recognize unfamiliar faces of a different race. They then had Caucasian participants practice identifying faces as individuals or practice categorizing faces by race. They found that after a few sessions of trying to see faces as individuals, participants showed significant improvements in their ability to recognize faces of a different race, as well as decreased implicit racial biases. Their finding suggests that implicit racial biases and poor recognition of other-race faces may both result from our tendencies to view people of a different race as categories instead of as individuals. However, their results also suggest that we can learn to view people as individuals—and this can help us overcome those biases.

My own research has shown that the influence of racial categories can even be overcome without training. In collaboration with Barbara Fredrickson of the University of North Carolina at Chapel Hill, I examined the influence of emotional states on the ability to recognize own-race and other-race faces. I had Caucasian participants view a short film to make them feel joyful, fearful, or neutral before they were given a facial recognition test.

The results were striking. Although participants who were made to feel neutral or fearful showed the typical pattern of being less able to recognize other-race faces, the participants who felt joyful showed absolutely no difference in own-race and other-race facial recognition. No training was necessary. Simply inducing a strong positive emotion before the recognition task eliminated the cross-race effect. It appeared as though positive emotions allowed people to view other-race faces less like categories and more like individuals.

In another study I examined whether positive emotions could decrease attention to racial categories. I had participants watch emotional film clips before doing a racial categorization task. My colleagues and I used facial EMG sensors to track how frequently participants expressed genuine smiles during the film clips. We found a direct relationship between how often someone smiled and the degree to which he or she focused on racial categories after watching the film clip. People who smiled frequently were significantly less aware of racial category markers a short time later, compared to those who did not smile frequently. Again, training wasn't necessary. The influence of racial categories could be overcome simply by smiling (frequently).

The links between smiling and racial categorization are just beginning to emerge, so at this point it is not entirely clear whether smiling by itself reduces racial categorizations or if there is something about the personalities of smiley people that also makes them pay less attention to racial categories. But some early evidence suggests smiling more can help reduce racial categorizations—and, in any event, smiling certainly doesn't hurt. The reduction in racial categorizations that I observed was not simply due to how happy or positive people felt throughout the experiment. In fact, smiling had a unique and independent effect on helping us to see each other as individuals.

The toll of prejudice

Making it a point to smile is a more pleasant way of countering racist tendencies than constantly trying to control one's behavior, for it turns out that trying to compensate for subtle behavioral biases can take a toll. When Northwestern University psychologist Jennifer Richeson and her colleagues had white and black college students interact with a white or a black interviewer, they found that after interacting with someone of a different race, the students had greater difficulty matching words to colors, a task that demands a lot of concentration. It seems that during the interracial interactions, students were trying to regulate their gestures, facial expressions, and behavior so that they wouldn't appear prejudiced—and this, not surprisingly, tired them out.

However, a simple change in perspective can ameliorate this kind of psychological toll. In another study—a favorite of mine—Richeson and her colleagues gave students different instructions before they engaged in an interracial interaction: Some participants were given no instructions. Others were told to try to avoid appearing prejudiced. The final group was just told to try to have a positive interaction. After the interaction, participants were tested on their ability to self-regulate—as with Richeson's other study, they were asked to match words to colors.

The results showed that the group given no instructions and the group told to try to avoid appearing prejudiced had great difficulty completing the task. The decrease in self-regulation—maintaining focus and concentration—was about the same for the two groups, which suggests that the group given no instructions may have been trying to avoid appearing prejudiced by default. However, the group that was simply told to have a positive interaction did not show a significant decrease in self-regulation. Treating the interaction as just an interaction, instead of worrying about the fact that it was an *interracial* interaction, was all that was required to avoid the usual depletion of mental resources.

Issues of racism and prejudice have been a source of struggle throughout history, and those issues certainly won't suddenly disappear if we all just smile a little more and try to be a little more optimistic. However, research is revealing the intimate connections between our feelings and expectations and our perceptions. When we encounter people of a different race, prelearned feelings and expectations can distort what we see. This line of research suggests that if we can greet those encounters with a smile and some genuine optimism, we may be able to better see people for who they really are. Smiles and positive perceptions may not make every prejudiced instance or impulse disappear, but every individual step forward is still a step closer to where we want to be.

Further Reading

Dovidio, J. F., K. Kawakami, and S. L. Gaertner. 2002. Implicit and explicit prejudice and interracial interaction. *Journal of Personality and Social Psychology* 82:62–68.

Ito, T. A., K. W. Chiao, P. G. Devine, T. S. Lorig, and J. T. Cacioppo. 2006. The influence of facial feedback on racial bias. *Psychological Science* 17:256–61.

Johnson, K. J., and B. L. Fredrickson. 2005. We all look the same to me: Positive emotions eliminate the own-race bias in face recognition. *Psychological Science* 16:875–81.

Trawalter, S., and J. A. Richeson. 2006. Regulatory focus and executive function after interracial interactions. *Journal of Experimental Psychology* 42:406–12.

Framed! Understanding Achievement Gaps

Rodolfo Mendoza-Denton

Nestled in the golden, foggy hills above the San Francisco Bay, a sculpted strand of DNA—1.2 billion times its actual size—sits in the courtyard of the Lawrence Hall of Science, a children's science museum at the University of California at Berkeley. My two-year-old son loves to climb all over this steel and plastic replica. I always chuckle at the ingenuity of the double helix as a play structure. The sculpture's giant proportions have always seemed appropriate to me: the discovery of DNA stands as one of the largest achievements in the annals of science.

But while few would dispute the extraordinary significance of DNA, there's a great deal of controversy over its role in determining the traits, behaviors, and skills that make each of us unique. This controversy becomes heated when it focuses on differences between groups, particularly when it comes to matters of intellect and academic achievement.

Last fall, one of the world's leading authorities on DNA chose to weigh in on this debate. James D. Watson, whose work identifying the architecture of DNA earned him a Nobel Prize in 1962, zeroed in on one of the most contentious aspects of this issue: the relationship between race and intelligence. In an interview with the *Times* (London), Watson said he was "inherently gloomy about the prospect of Africa." He went on,

> All our social policies are based on the fact that their intelligence is the same as ours—whereas all the testing says not

really. . . . People who have to deal with black employees find this not to be true.

Coming from someone else, it may be easy to dismiss such remarks as ludicrous. Yet when the codiscoverer of the double helix claims that people of one race are innately inferior to others, we cannot so easily put his comments aside. Can we learn anything from Watson's remarks?

The answer is yes. The lesson, however, lies not so much in his comments as in the assumptions hidden beneath them—assumptions about how and why groups perform differently from one another.

Watson was right that tests have indeed documented performance differences between racial and ethnic groups. In the United States we refer to these performance differences as "achievement gaps," with some groups (European and Asian Americans) consistently outperforming others (African Americans and Latinos) on standardized tests. These vexingly persistent gaps can be found across all levels of schooling, from National Assessment of Educational Progress (NAEP) scores in elementary school to SAT scores in high school and professional school tests, such as the GREs or LSATs.

But how we explain these differences plays a vital role in determining how we should address them. And it's here that Watson went so far astray. He traced these differences to genetics, when research shows that's just not true. They're really the products of prejudice.

Minding the gap

Watson's insinuations about the "inherently gloomy" prospects of Africans (and those of African descent) rest on two faulty assumptions. The first assumption is that intellectual ability, as surely as our hair color or the shape of our nose, is fixed and innate. The second is that our tools for measuring this ability—such as IQ tests or standardized exams—are faithful gauges of it, like thermometers are for temperature.

Watson's not alone in holding these beliefs; they're endorsed and actively used in U.S. culture in not-so-subtle ways. Our educational system tracks students according to "ability" levels. College and private high school admissions officers rely heavily on standardized test scores for their acceptance decisions. The inevitable conclusion from these assumptions—the one reached by Watson—is that if the tests point to group-based differences, those differences must be deeply ingrained in who we are. Indeed, a former president of Harvard, Lawrence Summers, once pointed to "issues of intrinsic aptitude" as the root of gender differences in math achievement.

Research in psychology, however, reveals that these tests are not immune to pernicious influences of prejudice. Psychologists Claude Steele and Joshua Aronson, for example, presented African and European American students with questions similar to those one might find on a standardized test. In the first condition, researchers told one group of test takers that the tests were designed to gauge their intellectual ability. In the second condition another group of students was given the same set of questions, but told that the researchers were interested only in the psychological processes involved in problem solving and would not assess the participants' ability.

The findings revealed that European American students scored just as well in the first condition as in the second. The African American students, however, performed worse than the European American students in the first condition, but performed as well as the European Americans in the second condition. In other words, the results were influenced by the way the tests were framed: the mere implication that the test was somehow tapping into "innate ability" was enough to disrupt the performance of the African American students—so much so that lifting this frame *erased* the differences in performance between the two groups.

This is not an isolated result. Other experiments show that something similar happens with gender: if men and women are told that a math test is designed to show gender differences, women reliably don't perform as well as men. But if the same test is presented as simply measuring math skills, those differences suddenly disappear. Given that these experiments use identical tests, the findings flatly contradict the notion that math ability is innately different between men and women. If that were the case, researchers wouldn't be able to eliminate test score differences simply by changing the way the test is presented to students.

How can a small change in the framing of a test lead to such dramatic differences in performance, and why are some minority students and women (at least in math) particularly sensitive to such frames? Steele and his colleagues have proposed that these situations activate "stereotype threat," whereby concern about being evaluated against, or perhaps even confirming, a negative stereotype makes people anxious and disrupts their ability to concentrate. Lift this threat and performance differences disappear.

This research suggests that stereotypes and prejudice can, in subtle but powerful ways, affect the data that is then used as "evidence" for the stereotypes' very existence. In other words, people such as James Watson point to poor test scores to affirm their own prejudices against African Americans. But these prejudices are contributing to the poor test scores, which then feed the prejudices. It is a vicious cycle, and it is played out across a wide variety of settings. Whether one is an older person learning how to operate a computer, a woman learning a new scientific procedure, or an African American taking the GRE, negative stereotypes can affect people in ways that seem to confirm these very stereotypes.

Identity matters

One of the frustrating paradoxes stemming from the phenomenon of stereotype threat lies in whom it most affects.

Imagine Vanessa, a young girl who from an early age is attracted to numbers and mathematical puzzles. Her interest and motivation translates into good grades in math in primary school, middle school, even through ninth or tenth grade. This is right around the age when young people begin to ask themselves, Who am I, what do I want to become? Vanessa begins to imagine herself as a mathematician, an engineer, even as a rocket scientist. She sets her sights on this identity, and slowly becomes more and more invested in her performance.

And yet she harbors doubts—echoed by peers, by some of her teachers and college counselors, and by an ever-present media promoting the idea that women are bad at math ("Math class is tough!" proclaimed a 1992 Barbie doll). These doubts come more and more to mind as she faces the natural progression of more difficult content in her area of study. She begins to notice that most of the professors in math and science at the universities she aspires to are men, and precisely because she's so invested in the domain, wonders all the more strongly if she can really hack it as a woman in the field. She feels all the more pressure and worry when taking qualifying exams and aptitude tests in math, and these intensified worries make her more likely to underperform.

Here lies the paradox—and the enduring cycle of inequality that stereotypes perpetuate. Research has shown that those who care the most about their performance—those who stake their identities in a given domain of achievement—are the ones who seem to be affected the most by the negative performance stereotypes in that domain. Being highly invested in what you do is often a key ingredient for success—yet when coupled with stereotype threat, this high level of self-identification can make one even *more* vulnerable to underperformance!

This insight helps explain in part why achievement differences between gender and ethnic/racial groups actually *increase* the higher one goes up the educational ladder—rather than decrease as one might expect if only those who are the most talented and most motivated populate the highest echelons of training. It also helps us see how this state of affairs further entrenches inequality, by disproportionately affecting those who, like Vanessa, are the most motivated to succeed and can challenge the stereotypes.

When motivated students like Vanessa stumble as classes get harder, it is easy to interpret their struggles as even stronger evidence for inherent differences in ability—the argument that everyone can paddle in the kiddie pool, but only the "real" swimmers can stay afloat in the ocean. But such a cynical view again misses the essential point that performance differences are so malleable—and so sensitive to stereotypes.

Psychologists Margaret Shih, Todd Pittinsky, and Nalini Ambady remind us of this in a dramatic study examining the math performance of Asian American women. One group of women was asked, before a math test, about the language they spoke at home, and how many generations of their family had lived in the United States—questions designed to prime or remind the students of their Asian American identity. Another group of women was first asked whether they preferred a co-ed or single-sex floor, and the reasons why they might prefer a single-sex floor—questions designed to prime this group with their gender identity.

Relative to a group of Asian American women who weren't asked about either identity, those women primed with their Asian American identity (positively stereotyped in math) performed better, and those primed with their gender identity (negatively stereotyped in math) performed worse. The fact that all the participants belonged to the same ostensible group underscores further the difficulty of tagging or assuming that we are able to discern ability on the basis of group membership. Not only is such ability sensitive to framing, but we each represent a unique constellation of different group memberships.

It can be very difficult, particularly within our culture, to appreciate the fact that intellectual performance can be so malleable. On Web sites and in magazines we are heavily drawn to IQ tests that promise to reveal how smart we "really" are. Yet we, along with Watson and Summers, fail to see how prejudices and negative expectations can help determine the results of these tests.

In my own research my colleagues and I have shown that notions about innate ability don't just hinder the performance of negatively stereotyped groups. It's worse than that. As in the case of the Asian American women who were reminded of alleged Asian math prowess, we found that stereotypes actually *boost* the performance of positively stereotyped groups. When people are led to believe that their abilities are fixed at birth, members of the positively stereotyped groups actually suffer less anxiety when they have to perform a task, because they're reassured that their group membership guarantees high ability. In other words, these stereotypes don't only perpetuate achievement gaps, they exacerbate them.

Think different

So what can we do about any of this? People often believe that, like our intellectual abilities, our prejudices are deeply entrenched in our genetic makeup and are unlikely to change. Yet again, research shows that our prejudices are quite malleable, and highly determined by social conditions.

When I was in school, and even as I completed my training as an educator, I would hear about the proverbial three *r*'s necessary for success: reading, 'riting, and 'rithmetic. This list belies the emphasis our society places on skills that can be measured and quantified by testing as the foundation for education. Research, however, is making increasingly clear the importance of a fourth *r*: relationships. Social relationships with people from many types of groups reduce prejudice, decrease social anxiety, and help generate positive experiences between groups. As such, these relationships are a key factor in shaping our prejudices, our reactions to stigma, and our academic achievement.

Studies suggest cross-group friendships are especially effective for improving relations between groups. For example, research by social psychologist Stefania Paolini and her colleagues shows that people who form friendships with a member of a different group are much more likely to hold positive attitudes toward that group, even among people who have experienced violence as a result of conflict between those groups.

But do these friendships actually affect people's attitudes? Or are people who hold more positive attitudes toward another group just more likely to befriend people from that group?

In my own research my colleagues and I found evidence for the actual impact of cross-race friendships. We took study participants who were initially racially prejudiced and facilitated friendships between them and people of another race. When we tracked their contact with other racial groups over time, the participants reported spending more time with people of other races. On the other hand, when we facilitated friendships between prejudiced participants and people of their own race, they reported no changes in contact with members of other groups. More recently, we have also found that when minority college students form cross-group friendships, they later feel a greater sense of belonging and satisfaction toward their university.

"Belonging" strikes many people as something that shouldn't matter for those hard, marketable school skills that are the stuff of standardized tests (think the traditional three *r*'s). Yet belonging matters, particularly when people have reason to suspect that due to their gender or their race, they are not accepted but merely tolerated as tokens.

Psychologists Gregory Walton and Geoffrey Cohen have recently applied this powerful insight to an educational program for African American college students at an elite, historically white university (recall how the most successful minority students are paradoxically the most vulnerable to stereotype threat effects). First-year students were told that all students, regardless of race, worry about their belonging on campus, and that these worries lessen with time. This "attributional retraining," though brief, helped assuage the students' belonging concerns as both normative and temporary—enough so that these students' GPA increased the following semester, while the GPA of African American students campus-wide who had not participated in the intervention decreased.

As many contributors to this anthology make clear, certain conditions raise the odds of creating environments that foster belonging and intergroup friendships. One of these conditions is the approval and support of authorities or institutions—for example, when universities take deliberate steps or adopt policies to make such contact possible, such as minority recruitment and social events designed to bring groups into contact with each other. Another condition is that members of different groups must enjoy equal status. That is, they should all feel equally valued and welcome in a social setting.

It is ironic that by tracking and ranking students, our schools not only reinforce notions of fixed ability; they make some students feel more valued than others, and communicate to students that those in authority do not condone contact across tracks or ability groupings.

In this way, practices such as ability tracking and ranking of students provide not one but two toxic ingredients that poison efforts at equality—a message that authorities endorse inequality, and a message that these inequalities are entrenched.

Luckily, the way we think about abilities is not permanently inscribed in our genes. It is within our power to change whether our nation's students see ability and intelligence as fixed or as something that can be improved with effort and hard work. In a powerful demonstration of this, psychologist Lisa Blackwell and her colleagues showed seventh-grade students how the brain grows and makes new connections when we learn, much like muscles becoming stronger. The exercise reminded the students that they can grow their intelligence. In subsequent testing, these students improved their math achievement across the difficult transition to junior high school, whereas another group of students, who had not been exposed to Blackwell's lessons, showed a decline.

In another study, which specifically focused on the gender achievement gap among adolescents, psychologist Catherine Good and her colleagues used a similar procedure to teach students that math skills are malleable, rather than fixed, and found that gender differences in math performance disappeared.

The good news is that we can reduce achievement gaps by teaching students that their academic ability is something they can work on, not something fixed by their DNA. The better news is that these interventions also seem to "lift all boats"—that is, everyone's achievement seems to benefit. And the best news may be that changing students' notions about fixed ability may itself lay the groundwork for reducing prejudice: if genes alone don't dictate how smart we are, it makes much less sense to believe that people's abilities depend on their race, ethnicity, or gender. We owe the tools of success to all of our students—not just to those whom we originally assume are going to succeed.

Outside Berkeley's Lawrence Hall of Science, the kids seem to agree that when climbing the DNA sculpture, the goal is to scale from one end to the other without touching the ground—a deceptively difficult feat given that the molecules twirl into a double helix. My young son can't pull it off yet, but I try to remind him that he wasn't born knowing how to climb. His missteps and falls don't mean he can't do it, just that he's learning how.

Further Reading

Mendoza-Denton, R., K. Kahn, and W. Chan. 2008. Can fixed views of ability boost performance in the context of favorable stereotypes? *Journal of Experimental Social Psychology* 44 (4): 1187–93.

Shih, M., T. L. Pittinsky, and N. Ambady. 1999. Stereotype susceptibility: Identity salience and shifts in quantitative performance. *Psychological Science* 10 (1): 80–83.

Steele, C. M., and J. Aronson. 1995. Stereotype threat and the intellectual test performance of African Americans. *Journal of Personality and Social Psychology* 69 (5): 797–811.

Walton, G. M., and G. L. Cohen. 2007. A question of belonging: Race, social fit, and achievement. *Journal of Personality and Social Psychology* 92 (1): 82–96.

When Racism Makes Us Sick

Eve Ekman and Jeremy Adam Smith

The Bayview Child Health Center is set back from Third Street, the chief thoroughfare in San Francisco's Bayview Hunters Point neighborhood. The street is lined with weathered pawnshops, check cashing outlets, small liquor and grocery stores, and bars. People in janitorial or business attire wait for public transportation to jobs downtown, while young men and women in baggy hip-hop gear hang out on the corners. Bayview Hunters Point has the highest concentration of families in San Francisco, but the Bayview Child Health Center is one of only a handful of pediatric clinics in this neighborhood.

Geographically isolated from the rest of San Francisco, a hot spot of violence and crime beset by toxic hazards from the neighborhood's shipbuilding industry, Bayview's largely African American community is struggling with a range of health problems. A 2004 health assessment by the city of San Francisco showed more heart failure, asthma, ambulance use, and low-birthweight babies in Bayview than in any other neighborhood in the city. The infant mortality rate in Bayview is the highest in all of California. But the poor health of Bayview's residents can't just be attributed to violence, poor air quality, lack of adequate medical care, and other physical hazards. They face another danger, one that's harder to quantify but perhaps no less harmful to their health: a chronic sense of powerlessness.

According to Nadine Burke, the medical director of the Bayview Child Health Center, to live in Bayview is to feel that you're not in control of your destiny, that your neighborhood is socially and economically weaker than surrounding communities.

"Power plays a huge role in my work," says Burke. "The messages here daily, all the time, are that you are not important and you have no power. When you are exposed day after day to poor living conditions, violence, and environmental injustice, you start to feel like you are not important and that your voice is not being heard. It's easy to feel like you don't have any power."

Research has linked such feelings of powerlessness to the kinds of health problems plaguing Bayview and many other communities around the world. This research shows that members of poor communities do not merely experience higher levels of violence; they are also more likely to have high blood pressure and frequent periods of increased heart rate, which contribute to a higher mortality rate. What's more, similar health problems have been shown to afflict the least powerful members of nonhuman primate species. Taken together, these and other findings suggest that the psychology of powerlessness can wreak havoc on people who sit low on the totem pole of any social structure.

"Poverty, and the poor health of the poor, is about much more than simply not having enough money," says Robert M. Sapolsky, professor of neurology at Stanford University. "It's about the stressors caused by a society that tolerates leaving so many of its members so far behind."

The agony of powerlessness

Sapolsky's research has been critical to uncovering the link between a sense of powerlessness and long-term health problems. In a series of studies, Sapolsky and colleagues found that baboons and other nonhuman primates low in their troop's social hierarchy face constant threats of petty violence and bullying from those above them. Their bodies respond to this stress by releasing the hormone cortisol, which gives the body an extra shot of energy, useful for short bursts of running or brawling—the so-called "fight or flight" response.

Moreover, Sapolsky argues, primates can turn on stress responses for purely psychological reasons. We remember previous threats and are plagued by the lingering fear that an attack could come at any time. In a primate who's been victimized in the past, even mild teasing can trigger a disproportionate stress response, such as a big shot of cortisol.

While cortisol might help primates confront or avoid immediate danger, it also increases blood pressure and blood sugar levels and weakens the immune system. As a result, continuously stressed-out primates are at much greater risk for diseases like hypertension, inflamed arteries, and insulin-resistant diabetes, and are more susceptible to problems with their immune and reproductive systems, as well as psychological disorders.

Humans may not face the threat of bloody violence as frequently as low-ranking baboons, but there are plenty of other ways we can feel disempowered. Worldwide, many studies have documented health disparities not only between the affluent and the poor but between the educated and the less educated, ethnic majorities and minorities, and other groups defined by power imbalances. In each case, members of the disadvantaged group have shorter life spans when compared with more powerful members of society. African American men, for example, are three to four times more likely to suffer strokes than white males. Black people are 30 percent more likely to die of cancer than white counterparts.

University of Michigan health behaviorist Arline Geronimus calls this the "weathering effect." Her research has documented that whites and blacks suffer hypertension at similar rates in their twenties, but blacks become sicker as they reach middle age. Similarly, the health of Latino immigrants tends to decline as they stay in America, even when their incomes increase. Geronimus argues that being a minority in a prejudiced society erodes one's health over time, as stresses and feelings of powerlessness accumulate.

Other research corroborates that this is not merely a function of environmental factors such as reduced access to health care or fresh food. When Harvard University's Ichiro Kawachi and colleagues analyzed a two-year slice of United States Census data, covering the deaths of almost four hundred thousand people, they focused on death by diseases with no known methods of prevention, treatment, or cure. Since no amount of money could buy freedom from these diseases, Kawachi assumed they might affect the powerful and powerless equally, without regard to income. Yet his results suggested otherwise: those who were lowest on the socioeconomic ladder were more likely to die—even though the ability to afford and access health care should have conferred no advantage. This suggested to Kawachi that feelings of power might have more of an impact on health than the material benefits associated with high income.

While Kawachi and his team used socioeconomic status as a marker of power, additional studies have revealed that health disparities can be rooted in feelings of discrimination. Vicki Mays, head of the University of California, Los Angeles, Department of Minority Health, argues that the perception of discrimination, as well as actual discrimination, can affect health just as strongly as sitting low on the socioeconomic ladder—with negative effects similar to what Sapolsky discovered among baboons. In a paper published in the 2007 *Annual Review of Psychology*, Mays and her coauthors review a range of evidence showing how perceived discrimination can activate the stress response, which translates into poor health for the victims of that discrimination. They report, for example, that compared to white American women, even African American women at the high end of the socioeconomic ladder are at higher risk of delivering low-birth-weight babies.

The stress of discrimination can also lead to risky behavior. When a team of epidemiologists in four American cities followed 3,300 black and white adults over a period of fifteen years, they discovered that 89 percent of blacks and 34 percent of whites reported that they had at some point felt discriminated against, and that those people were far more likely to engage in risky behavior such as drinking, smoking, and using illegal drugs.

The study also found that affluent and educated African Americans were more likely to report discrimination, while the reverse was true for whites. "That makes sense," says Luisa N. Borrell, lead author of the study and professor of epidemiology at Columbia University, "because African Americans who are poor are more likely to interact with people who are like them." But, she adds, affluent and highly educated African Americans are more likely to interact with white people, and it may be easier for them to feel singled out and thus be more sensitive to discrimination. Meanwhile, whites at lower income levels are often in the minority in their communities, and thus "get a double hit," feeling both poor and racially out of place.

Robert Sapolsky points out that low socioeconomic status and racial discrimination often work in tandem to create similar effects: "Both low socioeconomic status and racial discrimination tap into the same corrosive elements of psychosocial stress, namely lack of control and predictability, lack of coping outlets, and a static system that allows little room for optimism."

As the above research suggests, there's no simple formula to determine a person's level of power. It's not as easy as looking at any one factor, such as socioeconomic status, since feelings of personal power are shaped by a variety of subjectively experienced social and environmental forces. Social status, says Borrell, is in the eye of the beholder. "You're affected by looking at how far you are from the person next to you on the ladder," she says. "That's what creates stress."

To support this claim, Borrell points to research on the connection between income and personal happiness. A 2005 study by sociologists Glenn Firebaugh and Laura Tach found, for instance, that one's reported level of happiness in the United States depends less on one's income level than on how one's income compares to that earned by others in the same age group. Wide disparities in income can make people unhappy and stressed out, even when absolute incomes are relatively high.

Making it work

What can be done to address these health disparities? Borrell argues that there is only so much health professionals can do because the nature of the problem is political and societal. "It is something that is beyond the individual level. It needs to be addressed by the structure of the society," she says.

Harvard's Ichiro Kawachi agrees. He identifies a range of social policies that would be vital to promoting greater socioeconomic equality—and hence, better health. "Make an investment in education, for example, to give people a decent start in life," he says. "We can subsidize childcare, which is a major stress for low-income mothers, especially those who are single parents. We can expand unemployment insurance and expand access to health care. This is controversial only in the United States. Other societies view health care as a basic human right."

Kawachi says that to address health disparities, health professionals must go beyond their offices and clinics and enter the realms of social and government policy. "Health professionals ought to be among the first to go out there educating politicians, policymakers, and the public. There's only so much that medicine can do in putting bandages on people. The fundamental causes of health disparities in our society are failures in social policy."

This approach is embraced by Nadine Burke, whose work does not end at the Bayview Child Health Center. She participates in countless hearings, commissions, and planning sessions—and she is passionately committed to having a clinic that reflects the composition and needs of the surrounding community.

Burke emphasizes that to address health disparities that flow from differences in power, the community itself must set the agenda for change. She points to a "disconnect" between community members and the experts who study them, whose recommendations and treatments can exacerbate feelings of disempowerment. "The people who are involved in trying to find solutions are not necessarily the ones who are able to understand the problems in a way to properly address them," she says.

While Burke works to empower the community from her medical office, community activists like Espanola Jackson, sometimes known as "the mother of Bayview," are trying to change the balance of power in their neighborhood. For thirty years Jackson has led campaigns on issues from affordable housing to rallying for educational and professional training opportunities. Today, her primary fight is to clean up the toxic and radioactively contaminated shipyard in Bayview, which has sickened many Bayview Hunters Point residents. Jackson has seen a sense of powerlessness take its toll on the health of the community—and to her, the path to better health requires overcoming apathy and building political power. "Everyone must stop sitting back and letting the politicians tell them what is good for you," she says, noting that only twenty thousand people vote out of a pool of sixty thousand potential voters in Bayview. To improve the health of Bayview, says Jackson, "we need that power."

Burke knows the problems that arise from powerlessness will not end overnight. She sees her work as part of a long-term effort to build the power of the community. "It is important to have our clinic here in the Bayview, where there has been so much poverty and hopelessness," she says. "One of the most important things I can do is stay here and make it work. I think that sends a powerful message that we value our families enough to make sure that we bring the highest quality resources here, because they deserve it."

The Unhealthy Racist

Elizabeth Page-Gould

When we think about the victims of racism, we typically think of the immediate *targets* of racial prejudice: those who have suffered at the hand of discrimination and oppression. But new research has identified another, unlikely group of victims: the racists themselves.

In the urban metropolises of the United States and Canada, it is almost impossible to avoid talking to someone of another race. So imagine the toll it would take if every time you did, your body responded with an acute stress reaction: you experience a surge in stress hormones, and your heart pumps harder while your blood vessels constrict, inhibiting the flow of blood to your limbs and brain.

These types of bodily reactions are helpful in truly dangerous situations, but a number of recent studies have found that racially prejudiced people experience them even during benign social interactions with people of different races. This means that just navigating the supermarket, coffee shop, or modern workplace can be stressful for them. And if the racist person then has to go through this every single day, the repeated stress can become a chronic problem, which places him or her at heightened risk for disease in later life.

Harboring prejudice, it seems, may be bad for your health.

Challenge versus threat

The human body is incredibly adaptive to stressful situations. But our nervous system reacts very differently to stressful situations we perceive as challenges than to those we see as threats. It's a distinction that, in the long run, could mean the difference between life and death for people with racial prejudices.

Challenges incite a sequence of physiological responses that send more blood to our muscles and brains, enhancing our physical and cognitive performance. Threats, on the other hand, set off a physiological response that restricts our blood flow and releases the hormone cortisol, which breaks down muscle tissue and halts digestive processes so that the body can quickly muster the energy it needs to confront the threat. Over time, these responses wear down muscles, including the heart, and damage the immune system.

In other words, facing challenges is good for you; facing threats is not. And whether you perceive interracial interactions as a challenge or a threat may be the key to thriving in a multicultural society.

In one study, Wendy Berry Mendes, Jim Blascovich, and their colleagues invited European American men into the laboratory to engage in social interactions with African American men or with men of the same race as themselves. The participants were hooked up to equipment that measured the responses of their autonomic nervous system while they played the game Boggle with their white or black partners. When interacting with African American partners, the white men tended to respond as to a physiological threat, marked by diminished blood pumped through the heart and constriction of the circulatory system. However, European Americans who had positive experiences with African Americans in the past responded as though the game posed a challenge—increased blood pumped by the heart and dilation of the circulatory system.

This is not an isolated result. In a study with Rodolfo Mendoza-Denton and Linda Tropp, I randomly paired European American and Latino participants into same-race and cross-race pairs and had them disclose personal information to each other. At the beginning and end of the social interaction, participants provided saliva samples so we could measure their cortisol responses to the social interactions. Both Latino and European American participants who scored high on a measure of automatic prejudice—the degree to which you associate certain ethnic groups with the concepts of "bad" and "good"—had increases in cortisol during the friendly interaction with a cross-race partner, but produced less cortisol when interacting with a same-race partner. By comparison, participants who were low in prejudice were not stressed during either cross-race or same-race interactions.

In other words, prejudiced individuals perceived partners of a different race as a physical threat, even though they were in a safe laboratory setting and engaging in a task that was structured to build closeness between the participant pairs. This was true for both Latino and European American participants who were prejudiced. Imagine these same individuals trying to negotiate a racially diverse street scene or meeting at work.

In another study, Wendy Berry Mendes and her colleagues invited European Americans to take a survey over the Internet, measuring their levels of automatic prejudice against African Americans. These participants were then invited to a laboratory where either European Americans or African Americans evaluated them, as if in a job interview. Again, as in the study I did with my colleagues, cortisol spiked in the relatively prejudiced participants—and at the same time, their bodies released low levels of DHEA-S, a hormone that helps repair tissue damage caused by the taxing "flight or fight" response. In contrast, the more egalitarian participants—those who scored low in automatic prejudice—responded to the interracial interaction with greater increases in DHEA-S than cortisol, which suggests that they saw the evaluation more as a healthy challenge than as a threat.

A healthy society?

The bottom line is clear: harboring racist feelings in a multicultural society causes daily stress; this kind of stress can lead to chronic problems like cancer, hypertension, and type 2 diabetes. But interracial interactions are not inherently stressful. Less prejudiced people show markedly different physiological responses during interracial interactions. In all three of these studies, people who had positive attitudes about people of other races responded to interracial interactions in ways that were happy, healthy, and adaptive.

These positive attitudes can be learned; prejudiced people are not doomed to be that way forever. In my own study with Latino and

European American participants, we randomly assigned prejudiced participants—those who were measurably stressed out by simple cross-race conversations—to complete a series of friendship-building tasks over several weeks with people of a different race. Over the next several weeks, we watched cortisol levels diminish in prejudiced participants, a trend that lasted throughout the friendship meetings. Furthermore, in the ten days following their final friendship meeting, prejudiced participants who had made a cross-race friend in the lab sought out *more* daily interracial interactions afterward.

It's that simple: building friendships with people of other races seems to eliminate unhealthy stress responses, so that each new interaction can be greeted as a challenge instead of a threat. In a racially diverse society, those who feel comfortable with people of other races are at an advantage over those who do not. These results have profound implications for the way we design our neighborhoods and institutions; indeed, they suggest that race-mixing policies like affirmative action might be just as good for white people as for people of color. The future health of prejudiced people is not set in stone. If they're willing to take the first step and reach out to people of other groups in a friendly way, they may learn to thrive in a society that is increasingly diverse.

Further Reading

Blascovich, J., W. B. Mendes, S. B. Hunter, B. Lickel, and N. Kowai-Bell. 2001. Perceiver threat in social interactions with stigmatized others. *Journal of Personality and Social Psychology* 80:253–67.

Mendes, W. B., H. M. Gray, R. Mendoza-Denton, B. Major, and E. S. Epel. 2007. Why egalitarianism might be good for your health: Physiological thriving during stressful intergroup encounters. *Psychological Science* 18:991–98.

Page-Gould, E., R. Mendoza-Denton, and L. Tropp. 2008. With a little help from my cross-group friend: Reducing anxiety in intergroup contexts through cross-group friendship. *Journal of Personality and Social Psychology* 95 (5): 1080–94.

The Egalitarian Brain

David Amodio

As Susan Fiske and Kareem Johnson describe in the first two essays of this anthology, recent research in social neuroscience has revealed that prejudiced reactions are linked to rapidly activated structures in the brain that were developed long ago in our evolutionary history. Does this mean that racism is hardwired into our neural circuitry?

Far from it. One of the things we also know from neuroscience is that the human brain is built for flexibility in how we respond to our social environment. While normal responses that promote our safety and survival can lead to inadvertent prejudices, causing automatic reactions of alarm and distrust when we perceive someone from another racial group, there's more to the human brain than fear. We are also wired for cooperation and fairness. Research on the neuroscience of prejudice is simultaneously discovering the roots of egalitarianism—and revealing new ways in which the brain can overcome our initial fears and biases.

Blink of the eye

To understand prejudice and the brain, one must take the brain (and the mind) for what it really is: a survival machine. This may not be the most romantic way to describe the organ that poets like Emily Dickinson have exulted as "wider than the sky" and "deeper than the sea." But while our consciousness may be occupied with lofty thoughts, the brain is constantly working in the background like a personal assistant to take care of the details so we don't have to think about them consciously.

This of course includes the mundane, like breathing, regulating our heart rate, or automatically shifting our gaze toward threatening objects. It also includes our reflexive responses to a threat, when heart rate and respiration increase and blood is diverted to our larger muscles in preparation for fight or flight. These are normal responses that promote our safety and survival—and unfortunately, responses that can sometimes lead us to prejudice and discriminatory behavior.

However, the story doesn't end there. To understand how the brain overcomes initial responses to race, consider its evolutionary history. The basic machinery for gut reactions and snap judgments was present in the brains of our distant ancestors, and the same structures are still found in our brains today, primarily in the human subcortex. These relatively simple mechanisms for detecting "us" versus "them"—and for automatically treating "them" as a threat—are quite helpful for species living in basic societies that do not require cooperation with outside groups.

But with each step of our evolution, primate social networks grew in complexity, and the subtle demands of social interaction grew enormously. Alongside these changes came major increases in brain size. Humans now live in a multicultural society linked by neighborhoods, workplace and political hierarchies, states, nations, and global regions—and peaceful interdependence is now key to our survival. With these new societal complexities, the basic machinery of the mind that promoted the survival of our evolutionary ancestors becomes not so adaptive for social life in the twenty-first century.

During the process of evolution, the brain didn't simply get larger. It also developed completely new structures. In particular, the mammalian brain developed a neocortex—the outer "gray matter" layer of the brain—that grew atop the older subcortex (sometimes referred to as the "reptilian" brain). The neocortex provides a mechanism for fine-tuning and augmenting the functions of subcortical structures, like adding power steering and fuel injection to a car to enhance its performance.

To use another automobile analogy, imagine that the brain drives behavior like a person drives a car. Imagine that a teenager in a driver's education class is like the subcortex, and his expert instructor, sitting next to him with her own steering wheel and brake, is the neocortex. The student does well in most situations, but when it comes time to parallel park—an advanced maneuver—the instructor may have to take control of the wheel. The two drivers aren't in conflict—that is, they both have the goal of parking the car. But to perform this complex task effectively, the student needs the help of the instructor. In this way, the neocortex functions to take control of one's behaviors to override our immediate, but sometimes inappropriate, reactions to people from other groups.

The neuroscience of egalitarianism

How exactly does the neocortex keep our prejudices at bay? Most people would agree that nonprejudiced behavior involves treating people equally, regardless of their group membership. Indeed, what we really mean by "controlling prejudice" is sticking to one's goal in an interaction—whether it's asking for directions or evaluating a job candidate—without being influenced by race (or gender, or sexual orientation, and so forth).

While studies have shown that people are generally unable to deliberately turn down the intensity of a feeling or a stereotypic thought, people are quite effective at responding to those thoughts or feelings in a way that blocks the actual expression of bias. In other words, people can overcome racism by keeping their eyes on the prize. The brain cannot be antiracist, per se, because it never stops spotting differences and sorting people into categories. But it is progoal—and if the goal is to make judgments without regard to race, the brain can do that, though it may take a bit of effort and practice.

In a series of experiments, my colleagues and I studied the neural mechanisms that enable us to control behavior in the face of automatic prejudiced tendencies. In one study, we measured participants' brain activity while they completed a computer task that required them to override stereotyped tendencies. In the task, white participants were shown pictures of various handguns and hand tools. Their goal was to classify these objects as guns or tools by pressing buttons on the computer keyboard. But just before each gun or tool picture appeared, a face of either a white or black person flashed briefly on the screen. Given the stereotype that African Americans are dangerous, the momentary flash of a black face predisposes participants to expect to see a gun rather than a tool. This speeds up their response to guns and leads to more mistakes when a tool actually appears. In order to respond accurately on the task, participants need to override the influence of racial stereotypes. By measuring electrical changes in the brain as they completed this task, using electroencephalography (EEG), we hoped to shed light on the psychological processes involved in the control of prejudice.

We found that participants with positive attitudes toward black people showed greater activity in the left prefrontal cortex—a region associated with greater self-control—throughout the task. More interestingly, this increase in frontal cortical activity appeared to tune other regions of the brain to perceive the black and white faces differently. Finally, this tuning of perception helped participants to respond more carefully and accurately when categorizing the target pictures (guns and tools), and as a result, their responses were less influenced by racial stereotypes triggered by the faces.

In other words, less prejudiced people are more attentive to racial cues—and this helps them adjust their behaviors to respond without prejudice.

These results are similar to one of the studies Susan Fiske conducted, when she used functional magnetic resonance imaging (fMRI) to measure activity of the amygdala in response to black versus white faces while white subjects completed different types of tasks. When the task was to categorize faces according to their race, the researchers observed greater amygdala activity to black faces, suggesting a stronger emotional reaction toward blacks. However, when subjects had a specific task goal for which race was not relevant—for example, to search for a gray dot on the picture or to try to guess what type of vegetable the person in the picture preferred—the differential amygdala response to black versus white faces disappeared. These results provide even more support for the idea that people can override the effects of implicit racial bias by focusing their attention on their main, race-irrelevant task.

How does this play out in real life? Let's say that a white student approaches her black professor to ask about an assignment. If she remains focused on her questions—the point of the interaction—stereotypes related to the professor's race will be less likely to affect their interaction. Or imagine a person who just ran a marathon and is parched. When he walks up to the beverage stand for a bottle of water, his goal to find a drink may be so strong and focused that he doesn't even notice that the cashier is Arab, a racial group associated with terrorism in America. These goals are race-irrelevant, and so a strong focus on the goal prevents any stereotypes or prejudices from coming into play.

Course correction

But what happens if the marathon runner suddenly notices the Arab cashier's race as he begins to hand over his cash? After years of exposure to the Arab-as-terrorist stereotype, he could easily feel a gut-level sense of fear and momentarily freeze in place. His brain would then need to detect that this automatic response is inconsistent with the main goal (to buy a bottle of water), and would need to work extra hard to get the goal-driven behavior back on track.

A region of the frontal cortex called the anterior cingulate helps monitor the match between ongoing motor responses and one's behavioral intentions. If you were playing basketball and charging up to the basket, the anterior cingulate would be working hard to make sure your attention and movements were focused on taking the shot, undeterred by distractions. As the degree of mismatch between the intention and the response is detected—if, say, your arms aren't raising the basketball high enough—activity in the anterior cingulate rises, signaling to the prefrontal cortex (among other regions) that greater top-down control is needed to adjust your arms and make the shot.

Evidence for this process with regard to race has been shown in a series of studies. For example, on the gun/tool task described above, participants need to override the influence of African American stereotypes on some trials but not on others. Specifically, when a black face appears before a picture of a tool, the stereotyped tendency is to classify it as a gun, even though the goal is to classify it correctly as a tool. In a series of studies using EEG, we found that activity in the anterior cingulate increased when prejudice loomed and control was needed.

Moreover, subjects who had stronger anterior cingulate activity in response to race-biased conflict showed more accuracy in their behavior. That is, they were more effective in blocking the influence of the stereotype and focusing on the task at hand. In general, people who were more strongly motivated to respond without prejudice showed a greater neural sensitivity to the activation of racial stereotypes, and this is what helped them override stereotypes in their behavior.

Although the brain is often able to correct our responses before we make a mistake, there are certainly times when it fails. University of Wisconsin psychologist Patricia Devine describes many self-avowed egalitarians as being in the process of "breaking the prejudiced habit." Despite their beliefs and their best efforts, they occasionally slip up. However, other research shows that such slipups lead to renewed efforts to respond without prejudice, and with greater vigilance in situations where bias may occur. In particular, Purdue University psychologist Margo Monteith has shown that when a less prejudiced participant responds unintentionally with bias, she will be more attentive to racial cues and react more carefully in future situations. In other words, egalitarianism is a skill, and people can learn from their experiences to respond without prejudice.

My own research has shown that after such slipups, low-prejudice people experience heightened activity in the left frontal cortex—a region associated with greater controlled processing—and that this change in brain activity predicts their efforts to be less prejudiced in their future behaviors. In a sense, this work shows that a failure to act without prejudice can trigger stronger efforts to regulate one's behavior in future intergroup situations.

For example, let's say you make a quip to a colleague that comes off as unintentionally racist: "Hey, I'd always rather have a black guy on my basketball team!" Afterward, you feel guilty—and that's linked to neural processes that help you think twice before you speak in the future. For people concerned about nonconscious racist tendencies, this ability to learn from our mistakes gives us grounds for optimism.

Regulating prejudice

Is there any way to reduce that initial, automatic response to racial difference, the one that sends our primitive amygdala into high alert? Research on classical conditioning in rats suggests that once an emotional association is formed in subcortical circuitry, it is difficult, if not impossible, to unlearn.

However, the learning of new information may lessen the initial emotional response. Past research by me and others has shown the amygdala is more strongly activated when people look at faces of other racial groups. But more recent work suggests that new information about group memberships can change this pattern. For example, a study by Jay van Bavel of New York University showed pictures of black and white people to white participants. Participants were told they would be playing a game, and that some of the people would be on their team, and other people would be on an opposing team. When van Bavel scanned participants' brains while they looked at the faces, their amygdalae were more active while they viewed faces of the opposing team, regardless of race. That is, older negative associations with blacks seemed to be overwritten once participants learned that some blacks were on their own team, and thus were presumably friendly.

The implications of this study (and others like it) are powerful: we might be able to reduce automatic prejudice just by convincing people that they are all on the same team, be it a sports team, a company, a nation, or a planet. This idea, championed by psychologists Samuel Gaertner and John Dovidio, is called the ingroup identity model. Kurt Vonnegut had a similar idea in his 1959 book, *The Sirens of Titan,* in which an eccentric billionaire organizes an attack from Mars in order to unite the nations of Earth against a common enemy and, in doing so, fosters world peace.

Although it may be very difficult to eradicate prejudices, we can design institutions and interventions to change how we perceive people from other racial groups. The new psychology of racism suggests that simply suppressing prejudice—or trying to directly eradicate bias as it's activated in the brain—will not ultimately work. Instead, we have to let the amygdala do its job, and train ourselves to help the neocortex to do *its* job. We really don't have a choice—so many other aspects of life depend on our quick reactions and snap judgments, and it is a system that is designed to be relatively tamper-proof.

Reports of our "racist brains" have stolen headlines, depicting humans as victims to the unconscious prejudices lurking in the dark corners of our minds. But the fuller story portrays the human brain as being expertly equipped to overcome automatic prejudices and build positive social relationships. Through research on the neuroscience of prejudice, we have gained a better sense of how egalitarian intentions can succeed, as well as how they may fail. By knowing that biases work quickly to influence snap decisions, people can identify situations where prejudices may spring up, and then exert greater care in their actions. In this way, perhaps the egalitarian brain can help us build a more egalitarian society.

Further Reading

Amodio, D. M. 2008. The social neuroscience of intergroup relations. *European Review of Social Psychology* 19:1–54.

Amodio, D. M., P. G. Devine, and E. Harmon-Jones. 2007. A dynamic model of guilt: Implications for motivation and self-regulation in the context of prejudice. *Psychological Science* 18:524–30.

Amodio, D. M., P. G. Devine, and E. Harmon-Jones. 2008. Individual differences in the regulation of intergroup bias: The role of conflict monitoring and neural signals for control. *Journal of Personality and Social Psychology* 94:60–74.

PART II

OVERCOMING PREJUDICE

Introduction

The Editors

Based on the evidence presented so far in *Are We Born Racist?* it's tempting to answer the title's question with a simple yes. But though we seem to have deeply entrenched propensities to harbor racial prejudices, the research presented in this volume's first section also shows that we have innate skills to help us overcome these prejudices.

In the following section, our contributors explore efforts to translate science into practice and put these skills to use, bridging brain and society. They offer specific, concrete prescriptions for how the research covered in part 1 can and should inform—and sometimes already is informing—new programs and practices to overcome knee-jerk racism. Considered together, the contributions to this section take us on a journey from the building blocks of society—children and the family—to a broader assessment of how to develop healthier cross-race relationships across society at large, one conversation at a time.

But the science of racial prejudice makes clear that we can't simply expect individuals to overcome racism on their own; the problem is embedded too deeply—both in our individual psyches and across our entire society—for such an atomistic approach to work. Instead, as the essays in this section argue, we need to make systemic changes in the ways our educational curricula are developed, employees managed, and police officers trained. Each focuses on a different aspect of society—from schools to workplaces to law enforcement—but these pieces all center on a key insight: the fallacy of color blindness.

Advocates of color blindness maintain that skin color is a meaningless characteristic, and that everyone should be seen as the same, regardless of color. They argue that by acknowledging race—whether in public programs or private interactions—we create divisions between races. And in some ways, color blindness seems like the most important goal of the modern civil rights movement. Doesn't it follow naturally from Rev. Martin Luther King Jr.'s call for people to judge one another by the content of their character rather than the color of their skin? Yet the research covered in this book consistently shows that color blindness is an impractical, and even undesirable, goal; indeed, King himself never argued that we should pretend as though skin color doesn't exist. We're wired to see racial difference from a young age—as young as six months, contributor Allison Briscoe-Smith shows—and to deny that tendency gets us nowhere.

In fact, the essays in this section show that efforts to disregard race, though often well intentioned, can do more harm than good: they confuse kids just learning about racial difference, alienate students of color in the classroom, allow unspoken biases to fester in the workplace. We can't help but notice race, and trying to convince ourselves otherwise is likely to induce cognitive dissonance—or, worse, render ourselves blind to the socioeconomic inequities that still persist along racial lines. While research consistently finds that the only way to overcome deep-seated racial biases is to make ourselves more consciously aware of the associations we attach to perceived racial differences, the ideology of color blindness preaches obliviousness to race and racial disparities, in the hopes that they'll go away if we ignore them.

Despite the mass of scientific evidence demonstrating the futility of this approach, it has been gaining traction for years, especially in political and legal challenges to affirmative action. Those who subscribe to the ideology of color blindness interpret the civil rights movement as an attempt to banish race from the American consciousness altogether. And recently, this interpretation has been embraced by the Roberts Supreme Court, which has shown hostility to the idea of race-consciousness in any form, perhaps most notably in its 2007 decision invalidating programs in Kentucky and Seattle that used race as a factor in assigning children to particular public schools. "The way to stop discrimination on the basis of race is to stop discriminating on the basis of race," Chief Justice Roberts famously wrote in his majority opinion. In his concurring opinion, Justice Clarence Thomas cited Justice John Marshall Harlan's dissent in *Plessy v. Ferguson*, the case that upheld nineteenth-century segregation in the South. "My view of the Constitution is Justice Harlan's view in *Plessy*: 'Our Constitution is color-blind,'" wrote Thomas.

But what the color blindness camp overlooks—deliberately or not, it's hard to say—is that the roots of racial prejudice lie not in *whether* we perceive race but in *how* we perceive it. There's a crucial difference between seeing skin color as a marker of inferiority and seeing skin color at all: the latter is cognitively inevitable, the former is socially conditioned. A realistic goal for Americans—for all humanity—is not to train ourselves to be blind to the existence of racial differences. Rather, it's to become more attuned to how our brains and bodies respond to such differences, and why we've learned to respond as we do.

This isn't an easy process. It requires that we both be sensitive to cognitive reactions that take less than a second and scrutinize some deeply ingrained beliefs and assumptions. And it demands that our leaders—in the public and private sectors—create environments that encourage this kind of reflection. These are obviously challenging tasks, but research shows that we have the tools to tackle them. The essays in this next section demonstrate how we can put these tools to use.

How to Talk with Kids about Race

Allison Briscoe-Smith

Years before I became a child psychologist, I was a swim instructor and taught kindergarten. I remember working in the water one day with a four-year-old white girl when she started to rub my arm.

"Does it come off?" she asked.

"Does what come off?" I asked back.

"The black." She was rubbing her arm on mine to see if she could get some of my skin color on her.

Her mother, who had been sitting near us, gasped out loud. She turned to me, pale and embarrassed. "I don't know where she'd come up with such a thing," she said. "We never talk about . . . things like that." She pulled her daughter out of the water and ended the lesson, shushing the girl as they left.

As a teacher, I had heard these kinds of comments from children before—directed not just at me but at other kids or adults—then witnessed the crestfallen looks on their parents' faces. The parents would ask, "Where do kids get this stuff from—they can't even notice race yet, right?" or, "Does this mean my child will be a racist?" Or they would get defensive: "We don't teach them that stuff at home"; "We have plenty of friends of different races"; "We don't even talk about race, so how can they know what it is?"

In my work with children as a teacher and as a psychologist, I've found that scientific research can assuage many parents' fears. While there's no easy answer to the question, How do I raise a tolerant child? research does offer some constructive suggestions for how kids learn about race—and when and how to discuss it with them.

Let's start from the beginning: do kids even see or notice race? The answer is yes, they see and notice racial differences from a very young age, even in infancy. In fact, several studies by psychologists Phyllis Katz and Jennifer Kofkin have found that infants and very young children (from six to eighteen months) will look at the faces of people of a different race longer than they look at faces from their own racial group. This is how infants and toddlers commonly react to new information, and it suggests racial difference is visually salient to them. This means that kids are able to notice and pay attention to racial differences even before they can speak about them. Katz and Kofkin also found that, by the age of three, children will start choosing to play with people of their own race more than people of a different race.

While they may notice racial differences and even prefer playing with members of their own race, this doesn't mean that kids this young understand race in the same ways that adults do, nor does it mean they're burgeoning racists. For children under around the age of seven, race—or, rather, traits like skin color, language, and hair texture—are just signs that someone is in some way different from themselves, similar to gender, hair length, or weight. It's not unusual or unhealthy for kids to gravitate toward the familiar so early in life. Kids' views only become prejudiced when they start linking these traits to flaws in character or behavior. We adults are the ones who ascribe malice to simply noticing racial differences.

So in and of itself, recognizing racial difference is not a cause for alarm—quite the opposite, in fact. For years, studies have found that children who recognize these kinds of differences from an early age show a stronger general ability to identify subtle differences between categories like color, shape, and size—which, in turn, has been linked to higher performance on intelligence tests. Researcher Francis Aboud has found that children between the ages of four and seven who show this advanced ability to identify and categorize differences are actually less prejudiced. So parents, rest assured: when children notice and ask about racial differences, that is a normal and healthy stage of development.

Still, this can create a lot of work for parents as children ask questions and struggle to understand the racial world around them. On his blog Daddy Dialectic, Jeremy Adam Smith (one of the editors of this anthology) describes watching a basketball game with his four-year-old son. "Daddy," his son asked, "why do only black kids play basketball?" In that one question—which Smith says made him anxious—we can see a child noticing racial differences and trying to make sense of them. Smith's son may have also been implicitly asking, Since I'm not black, does that mean I shouldn't play? or, Does this mean that black kids are better at basketball than other children?

Now comes the tricky part: how do you answer those questions?

In fact, many parents have opted not to answer them. When psychologist Diane Hughes and colleagues reviewed research on how parents talk about race, they found that as many as half claim they don't talk about race at all with their children. These parents have, often with good intentions, embraced the ideal of color blindness. They assume that if they raise their children not to recognize racial differences, they'll prevent them from becoming racist.

The problem with this approach, however, is that we all do notice difference. When we abstain from discussing race with our kids, we may confuse them and implicitly send the message that it is bad or wrong to talk about racial differences. This may affect children of color as well as white children. For example, researchers Phillip Bowman and Cleopatra Howard found that when African American parents did not teach their children anything about race, those kids felt less prepared to handle racial discrimination, and in general they felt like they had less control over their lives or environments.

In addition, the "remain silent" strategy ignores the fact that we communicate how we feel and think about race in all kinds of ways to our children. Who we choose to befriend—and avoid—communicates volumes to our children. Do our children see us interacting across race lines? And if so, how do we act—anxious, friendly, concerned? Our children are watching us, making sense of the racial world throughout their development. We as parents have a great opportunity to begin shaping how our children think about race, to help create tolerant children early on.

Instead of trying to ignore race, research suggests that parents should be more proactive. They can tell their kids it's okay to recognize and talk about racial differences while still communicating that it's wrong to hold racial prejudices. When his son asked him why only black kids were playing basketball, Smith could have frozen in fear with concerns about "making his child racist" and simply ignored the question, shushed his son, or changed the subject, like many parents do. Instead, he gave his son a developmentally appropriate answer. "Well, I guess a lot of black kids like playing basketball," he said, and then posed a question of his own: "Do you want to play with them one day?" To which his son replied, "Yes!"

My own research with sixty-seven racially and ethnically diverse families with children under the age of seven indicates that talking and answering kids' questions about race may help them understand racial issues and become more tolerant. I found that parents who talked more about race had children who were better able to identify racism when they saw it and were also more likely to have positive views about ethnic minorities. This was true for both the white families and the families of color in my study.

Other researchers have made similar findings. A study done by Frances Aboud and Anna Beth Doyle took nine- to eleven-year-old children who held prejudiced attitudes toward ethnic minorities and placed them with other nine- to eleven-year-olds who held less biased beliefs. They asked the kids to talk for two minutes about some of the race-based beliefs they had endorsed earlier in the study. The results were remarkable: after these conversations, the high-prejudice kids demonstrated lower prejudice and more tolerant attitudes. Given this impact of a two-minute conversation with a peer, imagine what a childhood of conversations with parents could achieve.

While there is strong evidence suggesting parents should talk about race, researchers are still studying the best way to talk about it. For both white families and families of color, there is some evidence suggesting parents should avoid language that induces fear in their kids, because these kids won't know how to respond. For example, explaining to a child, "You know, people are going to be mean to you and treat you unfairly because you are X race," without providing coping skills, empathy for the child, or support, may actually cause more fear and bias toward others. However, following this kind of statement with, "But that doesn't mean we should be mean to others," and, "But those people don't really know how great you are and how special it is to be X," or, "And if that happens, you can come to me and I'll help you out," may actually provide the support and coping skills children need to handle such discrimination.

Other research by Bowman and Howard suggests that helping kids feel pride in their racial or ethnic identity helps boost their self-esteem—with the caveat that lessons of pride shouldn't undercut other groups. In other words, the message shouldn't be, "We're so much better and smarter than Ys," but rather it should support other groups, too: "You know, some Ys do things that way and that's great. We do things differently, and that's really nice, too." Teaching children about pride, and how to make sense of the differences around them, can actually be an act of teaching and supporting tolerance.

First and foremost, though, it seems that the simple act of just having these conversations about race can help. Given that research shows kids notice and are trying to make sense of race as early as six to eighteen months, these conversations can begin very early. To that end, it's important to make kids feel comfortable broaching the subject. That means parents should try to avoid making race seem like such a big or intimidating topic that kids believe it's off limits, and they should try not to make kids feel awkward or inappropriate for asking questions. One of the best ways for parents to do this is to practice talking about race—with friends, with each other, with colleagues—so they can reduce their own anxiety before discussing these issues with their kids. There are Web sites, chat rooms, and organizations out there to help parents get this kind of practice. For example, one place to start is the blog Anti-Racist Parent, where parents discuss their efforts to raise racially conscious kids.

So parents, next time you're on a playground and you hear your child say something that seems racially confused or even offensive, don't be embarrassed. Don't scold or shush. And don't end the conversation with, "We don't say things like that." Instead, you might want to try, "Hmm, why don't we talk about that some more?"

Further Reading

Aboud, F. 1988. *Children and prejudice.* New York: Blackwell.

Aboud, F., and A. Doyle. 1996. Does talk of race foster prejudice or tolerance in children? *Canadian Journal of Behavioural Science* 28 (3): 161–70.

Hughes, D., and L. Chen. 1999. The nature of parents' race-related communications to children: A developmental perspective. In *Child psychology: A handbook of contemporary issues,* 467–90. Ed. L. Balter and C. Tamis-LeMonda. Philadelphia: Psychology Press.

Katz, P., and J. Kofkin. 1997. Race, gender and young children. In *Developmental psychopathology: Perspectives on risk and disorder,* 51–74. Ed. S. S. Luthar, J. A. Burack, D. Cicchetti, and J. Weisz. New York: Cambridge University Press.

Promoting Tolerance and Equality in Schools

Jennifer Holladay

Today's students are the most racially tolerant generation our nation has ever seen. According to recent studies, they are more likely to have friends, or to date, across racial and ethnic lines than the generations that came before them. On the whole, they believe racism is wrong.

But we still have a long way to go. According to federal statistics, one in four students report being targets of racial or ethnic bias in a typical school year; one in ten say they've been called an offensive name at school; and, in urban schools, nearly 40 percent of students report seeing bigoted graffiti on campus. Disturbingly, schools are the third most common location for hate crimes.

So what can educators and schools do to help elevate levels of racial tolerance among students?

Since 1991 the Teaching Tolerance project of the Southern Poverty Law Center has served as a clearinghouse for promising practices and ideas. Through our semiannual magazine, curriculum tools, and Web site, we support more than five hundred thousand educators each year.

In 2007 Teaching Tolerance completed a two-year assessment of its programming. Our findings indicate that schools committed to reducing racial prejudice should focus their efforts in four areas: antiracist curricula, character education, cross-group contact, and school equity.

Using antiracist curricula

The use of multicultural materials is likely the most popular anti-bias intervention in schools. In a 2005 survey we conducted with the National Education Association and the Civil Rights Project (then at Harvard, now at the University of California at Los Angeles), the vast majority of educators reported that they use such materials in their classrooms.

Although multicultural materials offer great promise, particularly because they can be used in racially homogenous schools, researchers are somewhat split on their effectiveness. Even James Banks, coeditor of the *Handbook of Research on Multicultural Education*, is hesitant in his conclusions: "Curriculum interventions can help students to develop more positive racial attitudes, but . . . the effects of such interventions are not likely to be consistent."

In 2000 psychologists Frances Aboud and Sheri Levy examined findings from existing studies of multicultural education programs, dividing them into two groups: multicultural interventions (those that teach students about diverse groups) and antiracist/antibias interventions (those that specifically teach students about social problems, such as racism and xenophobia). Their analysis found that antiracist/antibias interventions appear to have a greater effect on reducing prejudice than multicultural ones.

In practice this means educators shouldn't simply teach about "great African Americans in history"; they must also teach about American racism and segregation and its effects on African American communities, past and present. Provided at no charge to educators, Teaching Tolerance's curricular materials employ this approach—and produce powerful effects on students' attitudes. For example, a 2005 assessment of our teaching kit, *The Children's March*, which focuses on how youths in Birmingham, Alabama, rose up in 1963 to bring an end to segregation, found that the unit increased students' levels of respect for less-accepted groups in the United States.

Other providers of top-notch, antiracist material include Rethinking Schools, which offers an exceptional collection of resources such as *The Line between Us: Teaching about the Border and Mexican Immigration*; Teaching for Change, which publishes *Beyond Heroes and Holidays: A Practical Guide to Multicultural, Anti-Racist Education*; and the Zinn Education Project, which provides materials that bring the work of Howard Zinn, author of the revolutionary *A People's History of the United States*, into middle- and upper-grade classrooms.

Deepening character education

Character education programs generally intend to help children develop moral decision-making skills; rarely are they designed with antibias aims in mind. However, these programs are relevant to prejudice reduction in an important way: longitudinal studies have shown that people with higher levels of moral development have lower levels of racial prejudice.

In most schools, character-education programs communicate a predetermined set of values to students, values such as "honesty," "responsibility," and "equality." Posters are hung on walls, worksheets are completed in homerooms, and morning announcements are populated with tidbits along the program's themes.

For antibias effects to take shape, however, schools must do more than tell students about a list of values or general principles, like "equality is important." Instead, we must prompt students to delve deeply into the nuances of these values and principles (e.g., "Does treating people equally mean treating everyone the same?").

To foster this kind of inquiry, several programs run structured dialogues that encourage students in grades seven and up to examine their own values and the values of others. Often designed to promote cross-group understanding, such programs can produce powerful results. For example, dialogue programs run on middle and high school campuses by Everyday Democracy (formerly the Study Circles Resource Center) have been proven to positively affect students' attitudes and to help create a shared understanding between white students and students of color about the realities of their schools' social climates. Similarly, an evaluation of the University of Michigan's dialogue-centered Program on Intergroup Relations showed that the program helped white students develop a greater sense of shared values and interests with people of color, while students of color perceived less racial divisiveness and reported more positive interactions with white students. These effects were sustained over three years.

Promoting cross-group contact

Wade Henderson of the Leadership Conference on Civil Rights once said, "Everyone talks about diversity, but no one talks about integration." He's right to draw the distinction. Integration between groups is far more likely to foster racial tolerance than is mere discussion of diversity's value.

For decades, researchers have studied social contact theory (or the contact hypothesis, as several authors refer to it in this volume), which asserts that when people interact across group lines, in properly designed activities, their prejudiced beliefs and behaviors will fall away. A 2000 analysis conducted by researchers Thomas Pettigrew and Linda Tropp, reviewing 203 individual studies and 90,000 subjects, confirmed the power of cross-group contact: in 94 percent of cases, face-to-face interaction between group members led to a reduction in prejudice.

Inside our nation's classrooms, social contact theory often comes alive through cooperative learning strategies, which generally involve breaking students into racially and ethnically diverse small groups. Group tasks require face-to-face contact and interdependence: an individual's success is dependent on the group's success. An analysis conducted in 1995 by psychologist Robert Slavin, reviewing dozens of existing studies, found that the use of cooperative learning strategies increases the likelihood that students will form friendships across racial and ethnic lines. Importantly, cooperative learning also boosts academic achievement among students of color.

Two factors currently make it hard for schools to adopt social contact strategies. First, as researchers at UCLA's Civil Rights Project have documented, school choice programs (e.g., vouchers), persistent residential segregation, and weakening legal mandates for integration are fueling the resegregation of American schools. As the twenty-first century gets underway, African American students are more likely to attend segregated schools than they were in 1968. Segregated schooling also is becoming the norm for Latino students—and the same goes for white children, who are the most racially isolated students in the nation. A decision by the U.S. Supreme Court in 2007, which ruled that school districts cannot integrate public schools by assigning students to particular schools on the basis of race, leaves districts with few options to pursue purposeful racial integration, and the resegregation of our schools is likely to escalate.

Second, the No Child Left Behind Act is placing extreme pressure on schools and teachers to "get test scores up." This pressure often translates into hours upon hours of classroom test prep and mandated use of scripted curricula, leaving little, if any, space for the practice of cooperative learning or any other form of instruction that's not seen as preparation for standardized tests.

Still, many schools remain diverse, and these schools should embrace programs, such as the Jigsaw Classroom and Teaching Tolerance's "Mix It Up at Lunch Day" initiative, that create opportunities for diverse students to cross group lines and get to know each other. Teachers, too, can use what flexibility they have to implement cooperative learning strategies in their classrooms.

Striving for school equity

If educators and policy makers truly want to reduce prejudice, they must work to provide all students with equal academic opportunities in schools. As Rodolfo Mendoza-Denton makes clear in his essay "Framed!" earlier in this volume, common practices like tracking or ability grouping can send strong, if unintended, messages to students about one group's supposed superiority or inferiority to others. If white students dominate a school's advanced placement courses and students of color tend to populate its remedial classes, for example, that reality can inform students' attitudes about race and racism in profound ways. In that case it's unlikely that antibias lesson plans can outweigh these attitudes and lived experiences.

One way we can begin to address such inequity is to better prepare teachers to work with a diverse student body. A recent report from Public Agenda found that, while a majority of new educators (76 percent) said teaching an ethnically diverse student body was covered by their training, only 39 percent said that training actually helped them in their classrooms. Teaching Tolerance's research supports these findings: we've consistently found that, while educators care deeply about their students' success, they often lack the support and tools necessary to create classroom communities that successfully serve students of color.

Recognizing this need, we partnered with the National Education Association, the American Association of Colleges of Teacher Education, and a national advisory board of renowned scholars to launch the Teaching Diverse Students Initiative (TDSi) in fall 2009. TDSi offers a research-based suite of professional development tools specifically designed to help educators understand how race and ethnicity should—and should not—inform their instructional practices, as well as how to create conditions in schools that enhance learning opportunities of diverse students.

TDSi is designed to be flexible and practical: educators can handpick material best suited to the immediate and ongoing needs of their school communities (e.g., an interrogation of color blindness or the commonly held "deficit" view of English Language Learners, in which educators tend to overlook these learners' strengths and assets). Further, examinations of prevalent racial attitudes like these always center on actual instruction and learning outcomes. (For example, if a teacher is mandated to use ability grouping, what should she do—if anything—when that grouping results in racial segregation? How might such grouping affect students' attitudes—and their academic performance?)

TDSi's resources are free and available exclusively online through the Teaching Tolerance Web site. It's our hope that this program will help equip educators to better serve the increasingly diverse students in their care.

Our schools have made tremendous progress in recent decades. But to reduce prejudice even further, and to provide a quality education to all our students, we must commit ourselves to the hard, uncomfortable work of examining the ways race and ethnicity play out in our classrooms and schools. We can realize our nation's promise of equality only through our demonstrated commitment to it—through our actions, day in and day out.

The Perils of Color Blindness

Dottie Blais

One Friday morning, right after teaching my third-period English class, I came face-to-face with my own racial and cultural prejudices.

I had just delivered a well-intentioned diatribe about the consequences of not doing homework. I couldn't understand why so many tenth graders were simply not reading their short-story assignments, or if they read them, were showing so little enthusiasm in class discussions.

One student, Julian, had remained in his seat after I dismissed the class. I sensed he had something to say.

Finally, he stood up and approached my desk. Julian was a tall, African American sixteen-year-old with intelligent eyes, one of my favorite students.

"You ought to quit trying to make us white," he said matter-of-factly. "All these stories you're making us read are by white people, about white people." His eyes pointed to the open textbook on my desk.

"Julian, I didn't select these stories on the basis of race," I said emphatically, stunned by the implications of racism. It was the unpardonable sin in a school where 70 percent of the students were ethnic minorities.

"Maybe you should have," he said, almost in a whisper, and left the classroom to silence.

I struggled to comprehend. His suggestion was inconceivable, especially since I had always prided myself on conducting "color-blind" classes.

I picked up the textbook and turned to the table of contents. Scanning titles and authors, I suddenly realized the awful truth. I hadn't deliberately eliminated writers of other ethnicities, but I hadn't deliberately included them either. Clearly, the assignments reflected my own unconscious cultural biases, and one student had the courage to say so.

Like many other well-intentioned educators, I had fully embraced the concept of color blindness. In theory, the approach sounds good. In a world where racial conflict has been such a problem, why wouldn't color blindness—with its attempt to remove race from the classroom altogether—be the perfect way to ensure that racism never rears its ugly head?

What I didn't realize until the incident with Julian is that color blindness has an ugliness of its own. The paradoxical truth, according to a 2004 study by Northwestern University psychologist Jennifer Richeson and her colleague Richard Nussbaum, is that the ideology of color blindness may cause more rather than less racial bias, a result that echoes other research findings.

How is that possible? To understand how a perspective intended to avoid racism in the classroom can actually encourage racism, educators must examine their own motives for employing the strategy, and several recent studies of color blindness can help them do so.

Teachers may believe, as I once did, that if they could somehow not see the race of their students, then they couldn't possibly be racist. However, a wave of new research has found that one of the first and most automatic ways people respond to others is to categorize them by race, even if they wish it were otherwise. In other words, achieving true color blindness is virtually impossible.

To deny what they inevitably see, many people employ what researchers call "strategic" color blindness, a strategy that requires maintaining an Orwellian perspective on race—that is, seeing and not seeing at the same time—to avoid the appearance of racism.

This is a recipe for disaster. In fact, a 2006 *Psychological Science* paper by Harvard University psychologist Michael Norton and colleagues reveals just how much this race-denying strategy can impair interracial communication. In one of their studies, white college-age students were randomly paired with a black or white partner (actually a confederate in the study, working with the researchers). Together they played a "political correctness" game: the white student had to examine thirty-two photographs that differed in three ways—by the gender of the subject, the color of the background, or the race of the subject. The black or white partner was given six of these thirty-two photos, and the object of the game was for the white "questioner" to identify the individual photos the "answerer" was looking at by asking as few yes-no questions about the photos as possible.

When whites were paired with whites, they were quick to ask about the race of the person depicted in each photo. But when they were paired with a black partner, they were significantly less likely to invoke race, even if their failure to do so meant scoring poorly in the game.

What's more, when whites tried to avoid mentioning race to black partners, their nonverbal behavior would change for the worse—they'd make less eye contact, for instance—and their communication in general would seem less friendly. Whites simply would not risk being perceived as racist, even though ignoring race in the game negatively affected their performance and their interactions with their partner.

A 2008 study published in the *Journal of Personality and Social Psychology* took Norton's research a step further, examining the effects that whites' attempts at color blindness had on black participants. Ironically, the negative nonverbal behaviors exhibited by "color-blind" whites were interpreted by blacks as signs of prejudice, making them suspicious of their partners. It is hardly surprising that racial tensions increased among participants.

What do these research findings mean for educators who truly want to avoid racism in the classroom? It seems clear that adopting a color-blind perspective, though often well intentioned and arguably a step in the right direction, does not actually combat prejudice among students and faculty; instead, it exacerbates racial divisions. We should abandon this practice. In its place, research suggests that a multicultural perspective, encouraging recognition and celebration of differences, is much more likely to reduce racial tensions and promote interracial communication. (See Jennifer Holladay's essay in this collection for more on the multicultural perspective.)

In retrospect, I can understand why Julian and other minority students in my English classroom were seething with resentment. By refusing to acknowledge that they were different from me, I elicited suspicion and discomfort. What's more, I insulted their racial and cultural identity. My color blindness signaled to them that they were invisible, somehow unworthy of my attention and my curriculum; my efforts were perceived as an attempt to "whiten" them.

Julian was right. If I had chosen to acknowledge race in my classroom, I could have intentionally developed a curriculum that was more inclusive. I could have provided an opportunity for students of all races to explore their own cultures through literature that extended and enriched the traditional canon. I could have used diversity as a learning opportunity for everyone in the classroom, including myself.

Julian taught me that if we choose not to acknowledge race in the classroom, students and teachers will rarely confront their own prejudices. Tolerance results from seeing commonalities. But unless we first see ourselves as different, we will never see ourselves as the same.

Overcoming Prejudice in the Workplace

Jennifer A. Chatman

In 2007 Intel released a print advertisement for its Core Duo Processor that caused an uproar. The ad, set in an office, featured six African Americans dressed like Olympic sprinters appearing to bow down to a Caucasian businessman. The general public found the ad to be "insensitive and insulting," according to Nancy Bhagat, Intel's vice president of corporate marketing. The company quickly terminated the ad campaign and several Intel executives made public apologies for it.

One employee at Intel, a former student of mine, told me the ad was a "big deal" for African American employees working there; understandably, many people at Intel and beyond were surprised that the company, with so many checks along the way, could have been so blind to the insulting nature of the ad.

I'll be honest: I wasn't that surprised. For years I've studied a variety of American workplaces, and I've found that prejudice and discrimination are alive and well within them. Though often unintentional, these prejudices aren't just a problem because of the emotional toll they take on victims, and they aren't just limited to discriminatory hiring practices. They can be insidious, creeping into many activities that affect the performance of individual employees, teams of employees, and the organization as a whole.

Take the example of Andrea, someone I met in the course of my research, who is a senior leader of a financial services firm. While serving on a task force with male peers, she noticed that her teammates rarely asked her opinion and didn't really listen to any of her ideas. After she mentioned this to one of her close colleagues, he confided in her that one of the team members, whom he knew, believed, as did the entire team, that she was appointed to the team for the sake of diversity rather than because of her distinct knowledge of the initiative that the task force was working on. The results: frustration, anxiety, and feelings of inadequacy.

Fortunately, my research has identified three steps managers—or, indeed, anyone in a leadership position—can take to reduce prejudice and its negative effects: publicizing strengths that go against stereotypes, emphasizing employees' shared fate, and selectively endorsing political correctness.

Publicizing strengths

As we can see in the case of Andrea, my research has found that if someone is a member of a group that has historically been underrepresented in a workplace—whether it's women, African Americans, or another group—coworkers will expect that person to perform poorly on tasks that have not typically been performed by members of his or her group.

This is true no matter how skilled the person actually is at that task. For example, coworkers expect women engineers to perform worse than men. Unfortunately, these expectations, whether conscious or not, are often self-fulfilling. Years, even generations, of prejudice weigh too heavily on the person and affect performance, even if he or she hasn't faced any explicit prejudice within the workplace.

However, I've also found that when coworkers are made aware of that person's expertise, the person's work no longer suffers. In fact, he or she excels, as does the rest of the team.

So how can colleagues learn of this person's expertise? One way is for the person to step up and advertise his or her own talents. But some people just aren't that outgoing and willing to hype their skills.

An effective alternative is for a manager to explicitly tell coworkers how skilled this person is, especially when the person joins a new work group. Can this seem patronizing? Perhaps. But my research suggests that it has a strong and positive impact—not just on this single person's performance, but on the performance of the entire group.

Indeed, this is exactly the approach that Andrea took on her next team assignment. She asked her boss to ensure that he made it clear what skills and experience she brought to the team—and, in fact, she found that the team was much more receptive to her point of view and regularly solicited her perspective. Perhaps as a result, the team finished its work more quickly, cheaply, and creatively than was expected. This may be because making clear Andrea's expertise gave her a confidence boost, but it also might have helped relieve the rest of the group of the discouraging notion that they'd have to "carry" a weak performer.

Emphasizing shared fate

Unfortunately, studies suggest that diverse workplaces often have lower productivity and camaraderie than homogenous workplaces. But managers can change this by encouraging people to recognize their common commitments to their organization rather than their individual, superficial differences.

In fact, I've found that when an organizational culture emphasizes employees' shared fate—the fact that they're all going to succeed or fail together as a group—diverse teams of employees are more productive and creative than homogenous ones. An organization's leaders can reinforce this sense of shared fate by, for example, rewarding entire teams rather than only individual members for their work. While such strategies may seem intuitive, surprisingly few organizations practice them consistently.

One organization I studied that does this well is IDEO, a product development company. IDEO's success rests on employees' ability to come up with creative and useable product ideas, and it intentionally hires diverse members—people with wildly different backgrounds and experiences—and then convinces them that their fate is intertwined. Indeed, one of the highest status roles in the organization is not coming up with a great idea, but offering an "assist"—that is, helping someone else develop their good idea into something great.

Endorsing political correctness (sometimes)

My research has also examined how people react to political correctness, which my colleagues and I define as censoring language that might be offensive to members of other demographic groups. Many managers are reluctant to advocate political correctness in the workplace, assuming that it stifles the free exchange of ideas.

In our research we either encouraged or discouraged teams to use politically correct language in their discussions, then we observed how these teams performed on a creativity task. In more homogenous teams, political correctness did seem to constrain creativity.

But in more diverse teams, we found that encouraging political correctness actually boosted creativity, while also promoting sensitivity to members' differences. It seemed that, though people are often anxious about cross-group interactions, political correctness provided clear ground rules for their conversations, promoting comfort and trust and enabling team members to focus their attention more completely on the task at hand.

Overall my research suggests that even after organizations achieve a diverse workplace, managers still need to take deliberate and consistent steps to address prejudice. But fortunately we have good reason to believe that these efforts pay off, fostering more just, harmonious, and productive workplaces.

Further Reading

Chatman, J., A. Boisnier, S. Spataro, C. Anderson, and J. Berdahl. 2008. Being distinctive versus being conspicuous: The effects of numeric status and sex-stereotyped tasks on individual performance in groups. *Organizational Behavior and Human Decision Processes* 107:141–60.

Chatman, J., and F. Flynn. 2001. The influence of demographic composition on the emergence and consequences of cooperative norms in groups. *Academy of Management Journal* 44 (5): 956–74.

Chatman, J., J. Polzer, S. Barsade, and M. Neale. 1998. Being different yet feeling similar: The influence of demographic composition and organizational culture on work processes and outcomes. *Administrative Science Quarterly* 43 (4): 749–80.

Flynn, F., J. Chatman, and S. Spataro. 2001. Getting to know you: The influence of personality on the impression formation and performance of demographically different people in organizations. *Administrative Science Quarterly* 46 (3): 414–42.

Goncalo, J., J. Chatman, M. Duguid, and J. Kennedy. Forthcoming. The unintended consequences of political correctness in mixed and same sex groups.

Policing Bias

Alex Dixon

In 1988 Sgt. Vernon Gudger was a rookie in Washington, D.C.'s Metropolitan Police Department. He was off duty one night at his mother's house when he noticed two black men about his own age attempting to steal the 1964 Chevrolet Corvair parked in his mom's backyard.

Gun in hand, Gudger ventured outside dressed in a baseball cap and a puffy red coat. He ordered the two men to the ground. Holding them at gunpoint, he told his mother to call the police to say he'd apprehended two suspects. But Gudger would soon learn that a neighbor had already called the police—reporting Gudger as the threat.

Three officers pulled up in a squad car behind Gudger and the men. Out came two black officers—Washington and Truesdale—and one white officer.

The officers drew their pistols, placing Sgt. Gudger—who himself is black—in a pickle.

"It was Washington and Truesdale who shouted 'Don't shoot, don't shoot! That's Gudge, that's Gudge! He's police!' " Gudger recalls. Gudger shouted back, "I'm police!" then followed his colleagues' directions, putting his gun down on the ground without turning toward the officers, despite his close proximity to the criminals.

After the suspects were cuffed, the white officer approached Gudger. "You're lucky they were there," he said, referring to the black officers, "or you'd be pushing up daisies."

Gudger was shocked—and he couldn't help but think that the incident would have played out differently had he been white. To many white officers, he contends, "every black person holding a gun is a suspect."

Indeed, incidents like this have elicited accusations of racism against police departments across the country. Over the last two decades, similar confrontations have ended in the deaths of innocents, like Omar Edwards, the African American New York City police officer shot and killed in May 2009 by a white colleague, and Amadou Diallo, the West African immigrant fired upon forty-one times by NYPD officers in 1999 after they mistook his wallet for a gun. As the furor over the arrest of Henry Louis Gates Jr. in July 2009 reminded us (Gates was arrested at his own home by police investigating a possible break-in) the role of racial bias in policing is still a highly contentious subject.

In response, cognitive scientists have set their sights on the psychology of police work, looking at how unconscious racial bias may inform the snap judgments officers have to make. Their results shed light on the deep cognitive roots of racial bias—and how these biases can complicate officers' decisions about when to pull the trigger. They're also informing a new wave of police trainings, attempts to reduce the odds that we see more cases like the Edwards or Diallo shootings.

Three blinks of an eye

Joshua Correll has been a leader in the psychological study of racial bias and policing. Correll, an assistant professor of psychology at the University of Chicago, first got interested in the subject after the Diallo shooting. Correll was a graduate student at the time; the incident inspired him to see whether the emerging science of racial bias might help explain the officers' reaction.

Correll conducted a study using a relatively simple computer simulation. In the simulation, a series of photographs flash on the screen. Some just depict different settings—the countryside, a city park, the facade of an apartment building. In other photos, black or white men are shown holding something in their hands—either harmless items, such as cell phones or wallets, or a gun. The participant is supposed to shoot the men with the guns and avoid shooting men without guns.

Participants have 850 milliseconds—about three blinks of an eye—to press a button labeled "shoot" or one labeled "don't shoot."

When Correll ran this study with white college students, he found that participants were more likely to shoot unarmed black men than unarmed white men and were less likely to shoot armed white men than armed black men. Their reaction times differed as well: they were quicker to shoot black men with guns than white men with guns, and they were slower to press the "don't shoot" button when an unarmed man was black than when he was white. These findings were even stronger for those participants who reported having more contact with blacks. When Correll ran this study with white and African American community members in Denver, he found that the two groups showed the same racial bias.

Tracie Keesee, the Denver Police Department's chief of research, was intrigued by Correll's study. The DPD had recently been at the center of controversy over the shooting of Paul Childs, a black, mentally disabled fifteen-year-old killed by Denver police officers.

"One of the questions that kept coming up in a lot of the community meetings was whether or not the Denver Police Department somehow trained their officers to focus on killing young African American males," says Keesee. "Of course our initial response was, 'No, of course not.' But we never really knew if our training may have been having some adverse or inverse impact on what we were trying to do."

Keesee contacted Correll, and the DPD became the first police department willing and able to work with him on his research.

For a 2007 study, Correll and his colleagues recruited roughly one hundred officers from the DPD, one hundred other residents of Denver, and one hundred or so additional police officers from across the country who had attended a Denver police conference. The participants were predominantly white, though they also included some black and Latino participants. All of the participants went through a simulation like the one in Correll's earlier study, where they had a split second to decide whether to shoot or not shoot white or black characters in a computer simulation, some armed with guns, some unarmed.

Similar to Correll's earlier study, the researchers found both police and community members were quicker to shoot blacks with guns than they were whites with guns, regardless of the participant's own race. And it took all participants more time to decide against shooting the nonthreatening photographs of blacks, the ones who were holding wallets or cell phones.

"What that suggests is that when they see a black target on the screen, the idea of injury or threat may pop into their minds," says Correll.

Research by many other psychologists and neuroscientists supports Correll's interpretation. As Susan Fiske describes in the first essay of this anthology, New York University neuroscientist Elizabeth Phelps and colleagues found that participants exhibited more activity in a brain region known as the amygdala when they saw black faces than when they saw white faces. Amygdala activity spikes when we feel threatened or afraid, suggesting that participants might have been experiencing fear or even aggression when they saw black faces.

And in a study lead by psychologist Keith Payne, a professor at the University of North Carolina at Chapel Hill, participants looking at a computer screen were asked to press one button when they saw a picture of a gun and another button when they saw a picture of a tool. But before they saw either kind of object, a photo of a black male or a white male flashed on the screen. The participants were faster at identifying the gun after first catching a fleeting glimpse of a black man's face—and they were more likely to mistake a tool for a gun after seeing a black face than a white one. (For more about the implications of this study, see Kareem Johnson's contribution to this anthology.)

Findings like these suggest to researchers that the brain is wired to respond quickly to possible threats, and in American culture at least, people may have been socially conditioned to see black male faces as one of these threats. This process can even affect people who consciously shun racial bias in any form, police officers who swear to uphold the law without prejudice, and people of color themselves. "Unless one is socially isolated, it is not possible to avoid acquiring evaluations of social groups," Phelps has written. "Yet such evaluations can affect behavior in subtle and often unintentional ways."

For police, researchers fear, this phenomenon might have fatal consequences, perhaps making them more likely to pull the trigger when they see a black suspect, especially when they think their own life could be on the line. This may help to explain why in the NYPD's 165-year history, there have been four incidents, including the Edwards case, where a white officer shot a black officer, but never has a black cop shot a white cop. During Sgt. Gudger's twenty-two years as a D.C. police officer, three black cops have been shot by white officers. Two were killed, and one is paralyzed.

Training the trigger finger

But it might not be so simple. In Correll's 2007 study, the time it took community members and police officers to shoot or not shoot seems to have been motivated by some unconscious bias. But when it came to their final decisions—that is, whether or not they actually pulled the trigger—police officers didn't appear to be influenced by race: they did not shoot unarmed blacks by mistake as often as ordinary people did, nor did they shoot them any more than they shot unarmed whites.

"That suggests that police are doing something," says Correll. "They demonstrate a pattern that we really don't fully see with any other group we've tested."

One reason police may have been more accurate, says Correll, could lie in the extensive firearm training programs police must undergo before they become officers. As they endure these trainings—from 360-degree computer simulations to trainings with paint guns—officers improve the way that their brains talk to their trigger fingers, learning when to show restraint.

"We're doing so much better at training police," says Lorie Fridell, a criminology professor at the University of South Florida. "So it could be that because use-of-force training has become as strong as it has, it's countered or offset the implicit racial biases of police."

Does this contradict the prevailing assumptions about cases like the Edwards and Diallo shootings, that they were the products of racial bias?

Not necessarily. Fridell maintains that though police firearm trainings may be effective in countering some racial bias, research shows that unconscious biases run deep in most people, and these trainings don't really get to the root of the problem.

"Even the best officers, because they're human, might practice biased policing," she says. National statistics seem to support Fridell's claim: according to an analysis by the Justice Policy Institute, for instance, white young people are significantly more likely to use or deal drugs than African American youth, yet black youth are arrested for drug offenses roughly twice as often.

Correll adds that though some of his research suggests police officers can keep unconscious racial biases in check, their ability to do so may be compromised when they're under stress, or at the end of a long workday. His current research is exploring how stress and fatigue may affect officers' decisions about whether to pull the trigger.

Fridell is now developing a training program with funding from the U.S. Department of Justice that is specifically tailored to combating unconscious racial bias. The program relies on two key strategies.

Fridell calls her first strategy "consciousness raising." The idea is to teach police officers that racial biases lurk beneath everyone's conscious minds, as the psychological research suggests. That way, police in the field will be more likely to catch themselves when their behavior may be unwittingly influenced by subtle biases. For instance, officers will engage in simulations in which they must make snap judgments about suspects, then step back and review how those judgments may have been swayed by the suspects' race.

Her second strategy relies on what's known in psychology as the contact hypothesis, and it's a phenomenon that goes far beyond policing. This idea holds that if someone has positive experiences with members of another racial or ethnic group, that person is less apt to be prejudiced. It may seem like common sense with respect to overt racism, but it also may affect unconscious bias. "The contact hypothesis has great implications," Fridell says. "Police have many factors to consider when hiring, and I certainly wouldn't want them to focus in on just one, but all things being equal, you might want the person who has had the diverse, positive experiences versus the one who has not." (The NYPD officer who shot and killed Omar Edwards grew up in a section of Long Island that is 85 percent white.)

A 2006 study by Florida State University psychologists Michelle Peruche and Ashby Plant supports this idea. Using a computer simulation very similar to Correll's, Peruche and Plant found that officers who reported having positive contact with black people were less likely to shoot unarmed black suspects than were officers who had negative attitudes toward black people.

Fridell's program is part of a broader movement to build on research like Correll's and bridge the gap between law enforcement and scientific research. The Policing Racial Bias Project, a program led by Jennifer Eberhardt, a psychology professor at Stanford University, has developed partnerships between police agencies and social psychologists, allowing agencies to participate in cutting-edge research into some of the unanswered questions about racial bias and its role in policing. In 2004 Eberhardt organized an unprecedented conference that brought together researchers and law enforcement officials from thirty-four agencies, spanning thirteen states. The conference was designed not only to help research inform policing, but to have policing inform social science, making its studies more relevant to real-world predicaments.

The Consortium for Police Leadership in Equity at the University of California, Los Angeles, is another such "matchmaker," helping police departments team up with world-class researchers. Already, nine police departments across North America have dedicated themselves to working with researchers—Correll, Keesee, and Eberhardt among them—as part of this consortium.

There are other signs that law enforcement is taking this research seriously. Following Edwards' death, the NYPD hired Correll to investigate the role that race may have played in the shooting. In 2007, after the *San Francisco Chronicle* reported that San Francisco arrested African Americans at a rate higher than any other California city, Fridell made twenty-eight recommendations to the San Francisco

Police Department about how to guard against racially biased policing. She was encouraged to find that the department was willing to follow many of these recommendations, such as allowing her to conduct bias training with department leaders. She now hopes that the SFPD will adopt the program she is developing for the Justice Department—and will be the first among many police departments to do so.

"The reality is that any department who hires human beings needs to be proactive in promoting fair and impartial policing," says Fridell. "That gets us away from pointing fingers at who is bad and who is good."

"People Understand Each Other by Talking"

Rodolfo Mendoza-Denton

On February 18, 2009, President Barack Obama's newly appointed attorney general, Eric Holder, delivered a now infamous speech on race relations. In that speech—a call for the nation to hold frank conversations about race—Holder admonished:

> In things racial we have always been and continue to be, in too many ways, essentially a nation of cowards. . . . We work with one another, lunch together and, when the event is at the workplace during work hours or shortly thereafter, we socialize with one another fairly well, irrespective of race. And yet even this interaction operates within certain limitations. We know, by "American instinct" and by learned behavior, that certain subjects are off limits and that to explore them risks, at best, embarrassment, and, at worst, the questioning of one's character.

Many people felt unfairly judged, and even attacked, by the attorney general's words. After all, barely a month prior to Holder's speech, the nation celebrated the inauguration of its first African American president—an election that for many symbolized its very courage to move toward a postracial society. Here was the nation's newly minted attorney general chastising its citizens for lack of boldness, right after they'd taken one of the most decisive steps against racism in American history.

Yet contemporary research on interracial communication does lend some support to Holder's claim that racial issues are especially difficult for Americans to discuss. Recent research by Sophie Trawalter and Jennifer Richeson, for example, has found that cross-race interactions provoke more anxiety than same-race interactions, and cross-race conversations are especially unnerving for white participants when they focus on the subject of race.

Given research findings like this one, we have to ask: how can a nation that embraces egalitarian ideals and recently elected its first African American president still have such a hard time addressing topics that made this election so historic in the first place?

New psychological research has shed light on an answer. Paradoxically, as it turns out, these egalitarian ideals that many of us cherish, and that helped elect Barack Obama as our president, are the very *same* ideals that lead to awkwardness and nervousness in interracial conversations and interactions. This new research helps us understand the pitfalls surrounding interracial interactions—and reveals a guiding set of principles to help make them more positive and rewarding.

Detours on the road to equality

In a recent study, Nicole Shelton, Jennifer Richeson, Jessica Salvatore, and Sophie Trawalter conducted a laboratory experiment in which black and white volunteers were asked to talk about race relations—precisely the kind of discussion that would make them uncomfortable, by Attorney General Holder's estimation. The researchers measured the white participants' implicit racial bias—that is, the strength of their knee-jerk, subconscious associations between black people and negative stereotypes. Following the conversation, the black participants provided their impressions of their white partners.

Surprisingly, the researchers found that the *less* biased the white partners were, the less their black partners liked them!

How do we make sense of these findings? Research by Jacquie Vorauer and Cory Turpie suggests that people who are low in prejudice want very much to show just how unprejudiced they are. As a result, they spend considerable energy monitoring their behavior—and can "choke" under the pressure of making a good impression. Thus, in Shelton and her colleagues' study, the less-biased participants may have actually been unable to pay full attention to the interaction itself because they were mentally busy making sure that they were in no way doing anything to make themselves come off as prejudiced. Ironically, their efforts to prove their egalitarianism may have made them poor conversation partners.

These findings echo research described in my essay "Framed!" in the first section of this anthology, which shows how a preoccupation with how one is perceived can undermine performance. For example, among minority group members, worries about confirming negative stereotypes and concerns about being the target of prejudice can undermine academic or job performance, trigger stress reactions, and lead people to avoid intergroup interactions altogether. Recently, psychologists Phil Goff, Claude Steele, and Paul Davies have proposed that during interracial interactions—and particularly when discussing race-related issues—whites may also experience the disruptive effects that stem from the threat of being stereotyped.

But what stereotype could whites feel threatened by? Goff and his colleagues speculate that it's the stereotype of being prejudiced—the "questioning of one's character" that Holder referred to. And just as black people's concerns about confirming a negative stereotype impair their performance, regardless of how skilled they actually are, white people's worries about confirming a stereotype of a "white racist" can impair their behavior in interracial interactions, regardless of their actual level of prejudice. The stereotypes are different, the situations are different—but the processes are the same for both blacks and whites.

Goff and his colleagues conducted an experiment that demonstrates this effect. The researchers had white participants arrange three chairs so that they could have a conversation with two black partners about racial profiling. The results showed that regardless of how prejudiced they were, the more participants were worried about confirming the white racist stereotype, the greater distance they put between themselves and their partners when arranging the chairs. It was their anxiety about seeming prejudiced, not their actual level of prejudice, that caused them to distance themselves from a person of a different race.

Why can't you read my mind?

Of course, even though behavior like that chair placement is motivated by an attempt to not seem racist, others can't usually just guess our good intentions. What is surprising, however, is that we behave as if the people we meet *should*, in fact, be mind readers—and we react negatively when they aren't.

For instance, Jacquie Vorauer has shown that people in cross-race encounters, and particularly people lower in prejudice, expect their anxiety and their egalitarian attitudes to be transparent to their partners. But, of course, these partners can't read their minds, they can only read their behavior—and as other research shows, this behavior is often tentative and awkward.

In related research, Nicole Shelton and Jennifer Richeson have shown that while both whites and blacks are actually interested in interracial interaction, both groups assume that the other group is *not* interested—and neither initiates interaction, based on this false assumption. This type of phenomenon, called pluralistic ignorance, arises when two or more parties act on incorrect assumptions about one another; it's like when a group of teens all end up drinking alcohol that nobody wants to consume only because everyone *believes* everybody else wants to. When Shelton and Richeson asked participants why they failed to interact with someone of a different race in a certain social situation, members of each group correctly said they avoided contact because they feared rejection, but they misattributed the other group's avoidance to lack of interest.

Unfortunately, these types of crossed signals can exacerbate an already stressful, delicate situation. In my own laboratory, my colleagues and I have found that when people feel their negative expectations for a cross-race interaction have been confirmed—when they think the other person is avoiding or behaving awkwardly toward them—they often react with strong feelings of anger and anxiety. This, in turn, makes it less likely they'll have the future interactions that might someday challenge their negative expectations.

A skill to learn

What hope do we have for heeding Attorney General Holder's advice to hold frank conversations about race if interracial communication is such a minefield? As it turns out, there is plenty of ground for optimism. But first, we need to reexamine our common assumption that people are either prejudiced or they're not. In most cases, the truth isn't so simple.

To understand why, consider classic research by social psychologist Patricia Devine, who has found that our minds are often the site of a battle between "automatic" and "controlled" components of prejudice. The automatic components are the associations that automatically come to mind as a result of how we've been socialized—a product of the world around us, from what we heard in our family growing up to the media we consume. These knee-jerk reactions can be useful—they're what make the word *snake* automatically activate a network of associations like "bite," "danger," and "run." But they're also what make a word like *Mexican* activate concepts such as "lazy" or "illiterate" for many people. Many of us fear that these fleeting associations reveal who we "really" are, like an X-ray into our true selves. By this reasoning, thinking stereotypic thoughts, or even acknowledging that we are aware of another person's race, can feel threatening, because these might signal that deep down, the answer to the question "Are you prejudiced?" is unambiguously yes.

However, as both Kareem Johnson and David Amodio argue in their contributions to this volume, these automatic associations exist alongside controlled, intentional behavior—efforts to reject and replace the fleeting, automatic stereotypes that sometimes cross our minds. Our intentional, egalitarian desire to not be prejudiced is a legitimate expression of our character. Research consistently shows that we can override our automatic associations through our behavior, and can even unlearn our automatic associations with enough practice. Thus we're not simply either egalitarian or prejudiced; egalitarianism is a learned skill.

In fact, in a follow-up study by Phil Goff and his colleagues, they found that when people approached conversations about race with the goal to learn from one another, rather than a goal to be evaluated positively, whites put their chairs significantly closer to their black partners than they had in the earlier study.

Interracial interactions can also be improved by acknowledging, rather than ignoring, group differences. Adopting a "color-blind" strategy in which one ignores group differences may seem like a good idea, especially if you're worried that even noticing race might mean that you're prejudiced. However, a wealth of research in both sociology and psychology shows that this strategy can actually *increase* racial bias. This is because race is almost impossible not to notice—as noted many times in this book, we may even be hardwired to do so—and thus the very act of monitoring whether we are noticing race makes it come to mind even more. By contrast, a "multicultural" perspective, in which group differences are acknowledged and celebrated, tends to be associated with less racial bias and more positive cross-race interactions for whites and blacks alike.

Contact, time, and patience

In light of all this research, the path to more positive interracial interactions may lie in three fairly simple prescriptions. Attorney General Holden highlighted the first—there must be opportunities for contact (a point lost in the furor over his use of the phrase "nation of cowards"). As argued throughout this anthology, the ability for people to spend time with each other can help improve interracial attitudes dramatically. In fact, this effect is so powerful that even the knowledge that one's *friends* have friends of a different race or ethnicity can improve one's own attitudes about that group—an effect labeled by Steve Wright and colleagues as the "extended contact effect." The extended contact effect has been shown to have deep effects. In their pluralistic ignorance study, for example, Shelton and Richeson found that when participants were told that their own best friends enjoyed interacting with members of another group, the participants were less likely to believe that members of this group were uninterested in intergroup interaction.

But the second and third prescriptions—time and patience—are just as critical. Without them, in fact, the positive effects of intergroup contact may not have time to take root. In my own research on cross-race friendship formation, for example, my colleagues and I found that those people who had a wealth of prior interracial contact (and were thus most comfortable in new interracial pairs) elicited the greatest stress reaction from partners who had been expecting a negative interaction. Why? Because challenges to our expectations, be they positive or negative, mean we cannot predict our environment, and this in itself is stressful.

Happily, over time, the cross-race interaction partners in our study achieved cross-race friendship—and that is precisely the point. Research suggests that frank conversations around race are likely to be saddled with negative expectations, so the beginnings of these conversations are likely to be rocky. We need patience for ourselves and others—not only in how to unlearn our biases and negative expectations, but also in how to learn from each other during our interracial interactions. And we need time—time to allow each other to be understood, time for us to relax, time for us to simply talk and get to know each other as human beings.

There is an old Mexican saying that goes, *hablando se entiende la gente*—"people understand each other by talking." My mom used to say this to my sister and me when we fought and gave each other the silent treatment, as a way to remind us that we were not taking each other's perspectives into account. The saying is so simple that it sounds almost obvious, but it helps to underscore that we need to find common ground between groups that typically view each other with unease. Recognizing this need, Attorney General Holder proposed

Black History Month as a time to hold conversations across the racial divide. But rather than grit our teeth and dig in our spurs for intense public debate, as the "nation of cowards" imagery might suggest, we can instead begin by acknowledging that such conversations are an everyday learning process, requiring nurturance and compassion.

Further Reading

Goff, P. A., C. M. Steele, and P. G. Davies. 2008. The space between us: Stereotype threat and distance in interracial contexts. *Journal of Personality and Social Psychology* 94 (1): 91–107.

Shelton, J. N., and J. A. Richeson. 2005. Intergroup contact and pluralistic ignorance. *Journal of Personality and Social Psychology* 88 (1): 91–107.

Shelton, J. N., J. A. Richeson, J. Salvatore, and S. Trawalter. 2005. Ironic effects of racial bias during interracial interactions. *Psychological Science* 16 (5): 397–402.

Trawalter, S., and J. A. Richeson. 2008. Let's talk about race, baby! When Whites' and Blacks' interracial contact experiences diverge. *Journal of Experimental Social Psychology* 44:1214–17.

Vorauer, J. D., and C. A. Turpie. 2004. Disruptive effects of vigilance on dominant group members' treatment of outgroup members: Choking versus shining under pressure. *Journal of Personality and Social Psychology* 87 (3): 384–99.

PART III

STRENGTHENING OUR MULTIRACIAL SOCIETY

Introduction

The Editors

In the previous section we saw how people can help each other to reflect upon their own biases—and thus limit the negative effects of prejudice. In the essays that follow, writers and scientists try to take the discussion about race to the next level, and describe worlds in which prejudice is, to some degree at least, overcome. From interracial marriage to the multiracial neighborhoods of California to the public fight for reconciliation in postapartheid South Africa, these writers reveal the imperfect ways in which individual human beings live in a multicultural world and embrace multiple identities.

Because these struggles can be so personal, it is important to remind ourselves that they are not new. As legal scholar Amy Chua convincingly argues in her 2007 book *Day of Empire*, peoples, nations, and empires have always struggled with racial and cultural difference—and many of them have been extremely successful. World-dominant powers since the days of the Great Persian Empire, established five hundred years before Christ, have strategically sought to integrate people of all colors and cultures into their institutions, sometimes adapting methods that seem to strangely anticipate our modern ones.

For example, ancient Rome actively offered pathways to citizenship and opportunity to people of all races, creeds, and languages—as opposed to excluding immigrants and refugees from citizenship, as some of today's U.S. pundits suggest we do. Interethnic marriage and sexual activity was commonplace, among the aristocracy as well as commoners (with Antony and Cleopatra being the most famous example). "Racism in the modern sense did not exist in Rome," writes Chua. "There is little evidence that Romans saw light skin as superior to dark skin, or vice versa." There were more slaves than citizens in the Roman Empire—which was no utopia—but they were not divided by skin color. This does not mean that people of the time didn't react negatively to racial differences, only that the Romans were able, for a time, to build an in-group identity that transcended those differences.

A thousand years later, Genghis Khan divided his warriors into interracial squads of ten "who were ordered to live with and defend one another as brothers," and he actively sought to promote non-Mongols to leadership, schemes that ultimately shaped the entire Mongol empire (and anticipated modern-day jigsaw classrooms, affirmative action, and cooperative learning, described by Jennifer Holladay, Jennifer Chatman, and Rodolfo Mendoza-Denton in their contributions to this volume).

Of course, none of the military and commercial empires Chua describes shared our contemporary ideals of equality and human rights. As she points out, tolerance in ancient Rome or Mongolia did not imply respect for other races and cultures. These empires were only more tolerant than their competitors, which, Chua argues, allowed them to flourish technologically, militarily, and economically; the rise of intolerance—of racial or religious chauvinism—often triggered their decline and fall. When the Roman emperor Augustus converted to Christianity and began persecuting the heretical Germans (criticized in racial terms as "drunkards," "wantonly brutal," and "rapacious"), allies and subjects turned against the empire and ultimately tore it down.

This history suggests that it is very much in our self-interest to foster tolerance—and to even imagine a society in which very different kinds of people "are permitted to coexist, participate, and rise in society," as Chua says the Persians, Romans, and Mongols tried to do long before modern-day Americans. These examples show us that there are ample precedents for multiracial societies—and that societies are often at their best when they embrace diversity and tolerance. In this history, we can also see the complex interplay between the prejudiced and egalitarian impulses that are present in every human mind and in every human society.

If there is a difference between the recent human past and today, it is that we have had the benefit of centuries of scientific progress. For hundreds of years, scientists and doctors have struggled to understand the relationships between the nervous system and our feelings and behaviors—and as a result, many citizens of the twenty-first century have gained a new appreciation for the ways in which we are more alike than different. The more we learn about what human beings are like under the skin—through research as well as through novels, memoir, and film—the more reluctant we become to dehumanize the Other.

This might be why, contrary to popular myth, interpersonal violence has actually *declined* over the course of recorded history. "Far from causing us to become more violent, something in modernity and its cultural institutions has made us nobler," writes psychologist Steven Pinker in an April 2009 essay for *Greater Good* magazine. "Today we are probably living in the most peaceful moment of our species' time on earth."

Inspired by the ideas of philosopher Peter Singer, journalist Robert Wright, and political scientist James Payne, Pinker suggests that evolution "bequeathed people a small kernel of empathy, which by default they apply only within a narrow circle of friends and relations." Over the centuries, however,

> People's moral circles have expanded to encompass larger and larger polities: the clan, the tribe, the nation, both sexes, other races, and even animals. The circle may have been pushed outward by expanding networks of reciprocity . . . but it might also be inflated by the inexorable logic of the Golden Rule: The more one knows and thinks about other living things, the harder it is to privilege one's own interests over theirs.

Today, the argument for multiculturalism goes beyond enlightened self-interest. In the following essays, writers and researchers reveal deeply personal struggles to expand their moral, emotional, and social circles to include people different from themselves, providing a glimpse into what a true multicultural America might look like—not a color-blind society, but one in which differences are recognized and embraced, not stigmatized. These essays suggest that we may never be able to ignore racial and ethnic differences, but we can change how we respond to them. A postracial society is not possible or desirable, but it is certainly within our power to build a postprejudice society.

Success Strategies for Interracial Couples

Anita Foeman and Terry Nance

For decades interracial couples faced prejudices, stigmas, and unfair treatment from most parts of American society, and academic research provided no exception. From the 1930s through the 1970s, research often seemed to reflect underlying, maybe unconscious, assumptions that interracial relationships must be fraught with difficulties; that those who would choose such a partnership must be doing so because they are acting out, or can't cut it with their own kind, or are trying to prove something; and that they and their partner must have identity and esteem issues. This research was usually conducted with interracial couples in crisis and in therapy, and not surprisingly, it showed that interracial couples are more prone to stress and divorce. It's a viewpoint that persists to this day.

As researchers of another generation, perspective, and motivation (we are both in long-term interracial marriages), we had other ideas about interracial relationships, and we put our cards on the table. We consciously sought out typical interracial couples in real-life situations to participate in interview research taking place in a variety of natural settings: church gatherings, dinner parties, playgrounds, and so on. Instead of trying to figure out what was wrong with these couples, our goal was to figure out what made their relationships work, day to day and year to year. Furthermore, we tried to identify ways to support people in cross-race relationships—mainly black and white couples—and offer the world insight as to why healthy, well-adjusted people might enjoy building a life with a person who happens to be from a different background.

What we discovered is that interracial couples do face considerable challenges—from a society that's quick to judge them, often right to their faces, and even from their own deep-seated racial prejudices. But we also found that many of these couples have developed effective strategies for dealing with these challenges, moving through four stages as their relationships deepen and evolve. They begin with a heightened awareness of the importance of race in society, and move on to find their own voice in the midst of the chatter around them. They learn to present themselves to the world in ways that protect their relationships, and, finally, they learn to pace their relationships for the long run. Understanding the journey of successful couples can be useful to people in interracial relationships, as well as to the educators, counselors, and friends who support them.

Race awareness

For the interracial couple, their relationship is rarely just about how they see one another. It's also about how society sees them together.

Most couples initially approach one another with fantasies that they will always feel completely compatible. The interracial couple, while perhaps sharing these feelings toward one another, may be acutely aware that society can see them as ill matched. The history of race and racism in America puts the couple in a position where complete strangers may feel that they have an investment in and a right to comment on the couple's relationship.

Couples we interviewed report that even during casual moments in their early courtship, friends, family members, and even people on the street made comments or approached them. Sometimes people were supportive, but in any case, the couple could not just explore their attraction in private. They had to confront the broader, political significance of their relationship early on, and they had to decide whether this relationship was worth all the discomfort and disapproval it might bring.

Such disapproval can come just from expressing cross-race attraction. A black man is likely to take heat for saying that he loves his partner's blond hair. A white man may be judged harshly for expressing attraction to his black partner's full lips. Some interracial partners we interviewed had been shamed and labeled for their feelings. Names like Uncle Tom, Oreo, "wigger," and "nigger lover" may be less socially acceptable in the age of Obama, but they still endure.

Depending on the individual's experience with past interracial friendships or romances, a potential partner may have to deal with disapproval in his or her own head. One interviewee said, "I used to refer to black people as niggers, and now I was dating one." Another person talked about how his beloved mother counseled him to "never bring a white girl into this house." If a particular racial or ethnic community or attitude shapes a person's identity, stepping out of that framework may challenge one's very sense of self. And these attitudes can undermine a potential relationship from the start: it can be very tough to reconcile the hostility a person feels toward another racial group with the attraction that person feels to a member of that group.

Plus, if a white person contemplates long-term involvement in an interracial relationship, he or she may have to address a possible loss of status. For some, especially white partners unconscious of the protection and privilege their race provides, this experience can be jolting. "Once you go black you never go back," said one white woman's father. "That's because we won't take you back!" Another white woman who happily told her dad she was dating an African American guy from the school basketball team reports that he just kept repeating, "Don't you want more than that for yourself?"

Yet another white respondent talked about being in a long-term relationship and finding himself shaken when his African American wife's nephew visited them. "We all went to the mall and he came with those low-hanging pants and a huge radio like the kids used to carry," he said. He panicked and thought, "I haven't just married her, I married her whole culture."

Finding space and voice

We found that if, after sizing up the challenges they face, couples remain committed to their relationship, the successful interracial couple is likely to look for their own safe space to find their voice together. They will want to explore each other's cultures and begin to build their own unique one. Their safe place is one from which they will strategize for dealing with threats that may come their way.

Despite the challenges of living in a society that sees race as a problem, many partners in interracial relationships describe this phase of discovery as exciting, interesting, and expansive. They explore everything from grooming (hair was mentioned a lot in our interviews!) to childhood family dynamics to differences in what they read, listen to, and do to unwind. One white participant talked about seeing *Jet*, the African American magazine, for the first time. A black participant talked about being introduced to the rock band Guns N' Roses. A white husband enjoys telling the story of how his African American wife wore heels on a camping trip.

Their experience with a person from another background gives them an opportunity to imagine new perspectives on the world and new relationship patterns that break out of tired old racial stereotypes. Several white male interviewees were taken by the independence, depth, and resilience of the African American women in their lives and saw this as opening the way to a new kind of partnership. Several black men talked about feeling freed of the negative profiles projected on to them in past relationships with black women.

Many people talked about never feeling that they fit neatly into society's image of white people or black people anyway. The possibility of a new option suddenly freed them to be themselves. One white woman said that she was relieved to be in an interracial relationship. She reports, "I never had the attitudes of other white people I know. I was always explaining my attitudes to people. Now, I don't have to explain. Everyone knows where I'm coming from, and they don't make a lot of racist comments around me."

Another partner says that she and her significant other are both "odd ducks" who were lucky to find one another and can now just be themselves and make their own rules. One African American woman actually felt more racially affirmed, saying that her white partner never tells her that she isn't "black enough."

Stepping outside of the issues of one's own culture can provide interesting affirmations. More than one black woman said that white men were less obsessed with light skin and "good hair" than black men ("After all, if that is what he wanted, he could date a white woman"). More than one white woman said that white men are more focused on the size-two woman than black men. One black man reported that throughout his young life many African Americans told him that he "talked like a white boy" and rejected him. White women never demeaned him in this way.

As partners in interracial couples share their stories and build their bonds, they explore the most sensitive issues of race that challenge society. A cross-race relationship creates a safe space for broad-ranging conversation on touchy racial issues; indeed, having them can build trust within couples.

We saw an example of this with the husband made uncomfortable by his wife's nephew with baggy pants. Many years later, the couple (still together) had to intensely process the husband's attitudes, directed at their own son, that reflected some deep-seated and negative attitudes that he held about black people and black culture. The wife says that her family and friends felt that her husband's attitudes were unfixable and unforgivable, arguing that she should leave the marriage. And yet, through their own conversations, she saw her husband evolve, and she said she looks forward to seeing that difficult process continue.

This experience holds a lesson for our entire society: the honest dialogue about the complexities of race that public officials like Attorney General Eric Holder call for is already happening in private, between interracial intimates. In fact, partners tell us that the insight they gain changes them in ways that would last even if the relationship does not. One of the husbands remarked, "Wow, if everyone truly loved a person who is black, racism would be intolerable."

Presenting themselves to the world

As couples work through many of these issues in private, they often become more confident addressing racial issues in public as well. They no longer simply respond to the ways others see them. They define their own experience, sometimes choosing to see themselves as exceptional or different but, ultimately, as existing in their own right and on their own terms, rather than as an inadequate version of a "real" couple.

In navigating tricky social situations together, they may devise a strategy to manage public challenges to their relationship. For example, an African American man decided with his partner that *he* would respond to black women who affront them in public. Others decide not to respond at all, or they agree on patent lines they will use. One woman says that when she is approached by someone who declares that people who date interracially don't like themselves, she responds, "If you would never date interracially, how could you know the motivations of someone who would?" Another self-described "nerd" says that he asks, "If I leave my partner, will you date me?"

They will make other practical decisions, like how to come out to others. Will each keep a photo of the other on a desk? Will they mention a partner's ethnic name as a hint? Couples will also decide which battles to fight. They may decide to live in an integrated neighborhood, or never again dine at a restaurant in the old neighborhood where "the maître d' pretended to have Tourette's syndrome and blurted out 'nigger'" periodically. They may decide to have only one partner attend a particular family function. The couple will learn to blend the elements of their separate cultures, maintaining what they most value (Irish dancing class is in), and shed elements that they find a difficult fit for them (dinner at Uncle Pat's is out). Some losses may be painful.

In maintaining a sense of family identity, couples often hold on to a stockpile of stories and images that serve as a touchstone for their family experience. For example, they may recall "the time when we stopped traffic in that small Southern town" or their surprise when someone they thought was disapproving showed up at a social event with their other-race partner.

Through it all, many describe their home as a "refuge," a "rock," or an "oasis." As they find a balance that works for them in public and private, they settle in for the long run.

Still, we have also found that many successful couples share a willingness to return to issues of race whenever, wherever, and however it is necessary. While the larger society may grow weary of talking about the topic or believe that "we did that in the 60s" or "I've had diversity training," interracial couples know that the process of relationship maintenance is ongoing, and if there is a racial issue confronting them, then they will "go there."

Ultimately, the successful interracial couple must be creative and adept in navigating their road. The happy couple enjoys the ride.

Supporting interracial relationships

The road can be rocky, but interracial couples shouldn't have to travel it alone; they often benefit from support from friends, mental health professionals, and others.

Friends and professionals who hope to support an interracial couple in their journey may want to begin by listening without judgment and asking questions. Supporters can ask the first questions of themselves. These questions can include: When is the last time I had a conversation of substance on any topic with a person of a different race? How many real friends do I have from another background? When was the last time I had a person of another race in my home? What kind of schools do my kids attend? or, Where do I worship, socialize, vacation, and call home?

If any or all of the answers reflect a person who is limited in their cross-race experiences, the ally can work on his or her own comfort and exposure to other groups. If supporters make efforts to expand their own circles of influence, it gives them credibility when they say they care. They certainly don't need to explain if they would not consider interracial relationships themselves, but reflecting on the reasons why may be important. If potential allies have ever asked interracial partners why they made their choice, they should expose their own relationship choices to the same scrutiny.

After serious self-reflection, a supporter can engage in dialogue with the interracial partner or couple, asking open questions that might cover what each values most in the relationship, to what extent their different backgrounds have shaped the evolution of their relationship, cultural styles that they each bring to the relationship that have enriched their life together, cultural elements that each has come to prize more as a result of the relationship, things that they've had to give up, and things that have been gained. Supporters should ask if there is anything they need to understand about the interracial couple or partner in order to be of help, and should also give interracial partners the opportunity and space to share thoughts or issues that come up for them during the course of such a conversation.

Having supportive friends, family, therapists, priests and ministers, neighbors, and coworkers makes all the difference in normalizing the relationship and helping the couple to feel like part of a loving community. As researchers, spouses, and Americans, we fundamentally believe that the insight interracial couples bring to their most important relationship can help this nation and world deal with one of its most vexing challenges and opportunities: how to love those different from us.

Further Reading

Foeman, A., and T. Nance. 2002. Building new cultures, reframing old images: Success strategies of interracial couples. *Howard Journal of Communications* 13 (3): 237–49.

Lawton, B. L., D. R. Arevalo, L. Brown, and A. Foeman. 2008. Coping strategies for interracial couples making decisions about children's education. *Iowa Journal of Communication* 40 (2): 155-79.

The Bicultural Advantage

Ross D. Parke, Scott Coltrane, and Thomas Schofield

Ellen and Tom Evans live in a nice suburban home in a safe neighborhood. They both have good jobs—Ellen's a nurse and Tom's a high school teacher—and have raised two teenagers who enjoy material comforts beyond Tom and Ellen's own Anglo-American middle-class origins. Their children, Mike and Lisa, are good kids, doing reasonably well in school, who aspire to go to college and become successful professionals.

Maria and Jose Lopez are first-generation Mexican immigrants. Maria works part time and Jose holds down two jobs—a daytime construction job and part-time evening work as a security guard—in order to support their four kids, who range in age from eight to sixteen. They live in a modest home on an urban street with lots of traffic and not much green space for recreation; they worry about the crime rate, the homeless, and the gangs in their part of the city.

So who enjoys a greater quality of life and tighter family bonds? The answer's not as simple as you might think.

When we take a closer look, we find that Ellen and Tom are struggling to balance work and family obligations as they try to maintain their comfortable suburban lifestyle. Their closest relatives live in another state, and they have few friends in the community. "It makes me sad that our kids don't see their grandparents regularly," says Ellen. "It's like they hardly know them." Although not divorced, Tom confesses that "we've talked about it on and off but so far we are holding things together." Ellen and Tom value family activities, but most of the time they do things separately from their children. "We each like to do our own thing, even though Mom and Dad want us to do stuff together," says fourteen-year-old Mike. "I'd rather spend time with my friends."

In contrast, the Lopez family enjoys a high level of support from their extended family and community. Their home is located close to their jobs, and they are part of a tightly knit Mexican American community. Many members of their extended family—grandparents, siblings, nieces, nephews, and cousins—live in the same neighborhood, and they frequently visit one another. "I feel real lucky that my family is close by," says Maria. "They help a lot with money when things get tight, and of course look out for the children. It's great." Unlike the Evanses, they do most things together—taking walks, going to movies, socializing, and attending church as a family.

The Evans and Lopez families both participated in a multiyear study we conducted of Mexican American and Anglo-American families. The differences between the Evanses and the Lopezes are emblematic of larger differences between their ethnic groups. Indeed, over the course of a decade of research, we've found that recently immigrated Mexican American families possess many of the strengths that we typically extol in American society—and which many Anglo-American families lack.

In general, Latino families are remaining strong and intact even though many parents have unstable jobs with low pay, limited benefits, and few opportunities for advancement—and despite the fact that roughly a third of Latino children live below the poverty line. They are facing great adversity—perhaps as much, if not more, than any other group in America. Yet divorce rates are lower for Latino families than they are for Anglo families with similar income and education levels.

In our research, we set out to discover why this is: what positive traits do Latino families—Mexican Americans in particular—bring to the United States, enabling them to persevere when so many other families flounder in a fast-paced, globalized world?

At a time when people are lamenting the death of the American family, we've found that many of the ideals of the American family are alive and well—and some of them are being imported from Mexico. Although our findings often run counter to conventional wisdom, we have found that Mexican immigrant families provide an inspiring model whose best features can be emulated by other American families.

All in the family

Decades of research have shown that kids do better in life if they grow up in a family environment that emphasizes closeness and support for one another. Studies have suggested that if children have strong, secure attachments to their parents, later in life they'll be better prepared to cope with stress, form healthy romantic relationships, and respond to the needs of their own children. Other research has indicated that mutual support among family members and a high level of family cohesiveness serve to benefit children's social and emotional development.

Researchers Ana Mari Cauce and Melanie Domenech Rodriguez argue that these strengths are defining features of Mexican American families. They maintain a commitment to familism, which emphasizes the importance of family closeness and getting along with, and contributing to, the well-being of the family.

This has proven true in our own observations. For instance, we've found that not only is the family unit strengthened by this commitment to familism in Mexican American families, but sibling relationships are stronger as well. Mexican American children in our studies receive more nurturance and social support from siblings and admire their siblings more than comparable Anglo-American children. Older siblings recognize that they are models for their younger brothers and sisters. "There are two things that stopped me [from joining a gang]," says Mario, a Mexican American teenager. "Mostly it was because of my parents and my brothers. They watch me in everything I do. Then they try to do better than me. So if I try to get good grades then they try to get good grades. I'm cool with it."

Latino families also often draw upon an extended network of

loved ones to provide necessary support, care, and guidance for their children. In addition to biological parents, adult caregivers may include grandparents, aunts, uncles, and godparents, as well as older children who care for their younger siblings. "I think he'll [stay on the good path]," says a Mexican American mother about her son in a study by University of California, Santa Cruz, researcher Margarita Azmita, "because not just myself but the rest of the family is not letting up on him. They talk to him constantly, reinforce the positive things he does do." This mother's intuition is backed up by research: Moncrieff Cochran of Cornell University has found that extended networks of family support promote positive traits in adolescents, such as better school performance and attendance, and more positive social behavior.

The collective responsibility for children in Mexican culture is captured in the practice of *el compadrazgo*, which involves special friends who become godparents of children. In a survey of Mexican American families in Los Angeles, anthropologist Susan Keefe found that 88 percent reported having compadres, with 80 percent of them living in the nearby area. Compadres serve as role models for their godchildren and are their potential surrogate parents as well. Moreover, most expect to provide food, shelter, and monetary—as well as emotional—support if needed.

Although Anglo-American families value their ties to extended family, too, their contact with family is less frequent and also requires travel across longer distances, due to the greater mobility of their families. Latino families, in contrast, are more likely to value living near family and do, in fact, choose to live in closer physical proximity to their relatives. Manuel, a Mexican American dad in our study, was amazed that many people do not always appreciate the importance of staying close to family.

"I don't know how people throw their parents into places where they die," says Manuel. "We kept our parents with the household until they died. And they are never too old for you to listen to. You take care of your parents. You are going to support them because they raised you when you were a kid, and we never forget that." Given such attitudes, it should not come as a surprise that we've found Mexican American households are less socially isolated than their Anglo-American counterparts.

Respect for elders

In our observations of recently immigrated Mexican American families, we find that parents teach their children to become *bien educado*, which means they develop strong social skills without being disrespectful toward adults. For example, children are taught to use the formal you (*usted*) rather than the informal you (*tú*) in their conversations with adults. Parents also encourage their children to participate in family rituals—another way to instill this kind of respect for their family's elders.

Politeness is not the only goal of these kinds of practices; they also serve to nurture a sense of obligation to one's family. We find that Mexican American children in our study more often help their parents in household chores than do Anglo-American children—indeed, Mexican American parents are more likely to require that their children complete such tasks before playing. They are also expected to play a more consistent and vigilant role in supervising their younger siblings than their Anglo-American counterparts. This ethic would seem to translate into other social spheres: when George Knight and his colleagues at Arizona State University watched different groups of children playing games together, they found that Latino children exhibit higher rates of cooperation and less competition than Anglo-American children.

Since immigrant children and adolescents often learn the language, customs, and norms of the host culture faster than their parents, they sometimes serve as cultural brokers or translators on behalf of their parents or grandparents. As Raymond Buriel of Pomona College has found, children and adolescents play a major role in helping their parents negotiate the legal maze, medical systems, and even workplace bureaucracies. These activities not only help the family unit thrive but provide children with opportunities to develop personal responsibility, self-esteem, and autonomy. (At the same time, children can use their linguistic edge to take advantage of their parents. One second-generation thirteen-year-old Mexican American boy admitted to Harvard researchers Carola and Marcelo Suarez-Orozco that he had told his parents that the F on his report card stood for "fabulous"!)

Father figures

A great deal of scientific research has shown how important it is for both parents to be involved in raising their children. In previous studies, we (Parke and Coltrane) have found that children of involved dads are better adjusted than those with less-involved fathers. For example, they seem to have less depression and better social relationships with their peers.

Although popular culture tends to describe Mexican immigrant men as more macho and less involved in family life, we are finding that these aren't accurate stereotypes. Instead, we find that Mexican-born men often exhibit higher levels of commitment to family and spend more time interacting with their children in nurturing and emotional ways than do Anglo fathers (or more acculturated Latino fathers, for that matter). Compared to Anglo-American fathers, Mexican-born fathers in our study are somewhat more involved with their children, interacting in traditional masculine activities like coaching soccer or playing games, but also in more routine activities like supervising children or taking them shopping. Yvonne Caldera at Texas Tech University has found similar involvement among U.S.-born Mexican American fathers. "I do the housework but he also helps," says one Mexican American mother in Caldera's study. "I go to work at six in the evening—and from there on he's in charge of the house. He feeds the children dinner and he leaves the kitchen clean for me."

We attribute these higher levels of father involvement to the influence of familism in Mexican American families—that is, high levels of family cohesion, cooperation, and reciprocity encourage these men to focus on the health and well-being of their children, and to interact with them in warm and intimate ways. For example, the emphasis on eating meals together and participating in family activities on weekends provides lots of opportunities for fathers to interact with their children and monitor their activities. In fact, in our study, families that engaged in rituals together—shared outings, mealtimes, and weekend activities—were the ones in which parents more closely monitored their children. Other research shows that this monitoring is associated with fewer social problems for adolescents, such as delinquency and substance abuse.

The bicultural family

We often assume that immigration is a one-way process: people from other countries come to the United States to settle and work, and they routinely adopt the values, customs, and practices of the host country. This is an oversimplified view that ignores the mutual influences between cultural groups.

We have found that many members of families that have recently immigrated from Mexico embrace a bicultural orientation, picking and choosing traits and practices of the dominant U.S. culture that help them to survive and thrive, while still retaining distinctive aspects of their culture of origin. "We must adjust to the way of life here," observes Juan, a Mexican American father, "but it shouldn't affect [my children] speaking Spanish and learning it correctly."

Rather than a liability, a bicultural orientation comes with clear benefits. Both children and adults who straddle the cultural fence, in fact, have better physical and psychological health, including higher expectations and feelings of positive self-worth, according to Raymond Buriel.

Other scholars, including Jeannie Gutierrez of the Erikson Institute and Arnold Sameroff of the University of Michigan, find that bicultural mothers have a more sophisticated understanding of children's development than do Mexican American mothers who have become more integrated into American culture, while their children do better both academically and socially.

This is a key lesson that immigrant families might have to teach many nonimmigrant families: understanding other cultural practices, values, and beliefs may not only increase our tolerance and acceptance of differences—a lesson that is valuable for both children and adults in our increasingly multicultural society—but also contribute to our own well-being. Just as biculturalism is beneficial for Mexican American immigrants, it could benefit other families as well.

Mexican American families teach us that a renewed commitment to the centrality of family in our lives could provide significant social and emotional benefits, such as greater buffers against the stresses, strains, and sorrows of everyday life. Our own research suggests that the negative impacts of financial hardship and other stressful life circumstances on children are reduced in Mexican American families with high levels of family cohesion. From the model provided by Mexican American families, we could also learn a lot about how community identity and community responsibility promote the welfare of children. The presence of many eyes and ears to monitor our children in public places would be a welcome change for many overextended, overworked, and underavailable families.

We are not advocating a Pollyannaish return to a nostalgic vision of family life. We recognize that some of the hierarchical aspects of traditional Latino family life are not desirable for many Anglo families—nor for many Latino families. Modern women who are active members of the workforce are unlikely to welcome a return to patriarchal practices where husbands rule and women obey. At the same time, in fairness, the stereotype of the patriarchal Mexican American family is outdated. Most research shows that Mexican American couples have moved toward a more equal balance of power and rights between spouses. Our research shows that Mexican American couples share parenting and housework in response to the same sorts of practical pressures faced by other couples.

By no means should we give up the positive gains toward more equal family roles for men and women in the United States that have been achieved over the last thirty years. But we would like to graft on to this newly emerging model some of the passionate commitment to family and community that characterizes the Mexican immigrant family. A new family form could emerge that is a fusion of our egalitarian family model and one that is better anchored by extended kin, neighbors, and communities committed to the common good of our children. Such a synthesis may be possible, and it is certainly desirable.

Further Reading

Adams, M., S. Coltrane, and R. D. Parke. 2007. Cross-ethnic applicability of the gender-based attitudes toward marriage and child rearing scales. *Sex Roles* 56 (5–6): 325–39.

Behnke, A. O., S. M. MacDermid, S. Coltrane, R. D. Parke, S. Duffy, and K. F. Widaman. 2008. Family cohesion in the lives of Mexican American and European American parents. *Journal of Marriage and the Family* 70 (4): 1045–59.

Coltrane, S., R. D. Parke, and M. Adams. 2004. Complexity of father involvement in low-income Mexican American families." *Family Relations* 53 (2): 179–89.

Parke, R. D., S. Coltrane, S. Duffy, R. Buriel, J. Powers, S. French, and K. F. Widaman. 2004. Economic stress, parenting and child adjustment in Mexican American and European American families. *Child Development* 75 (6): 1–25.

Pinto, K. M., and S. Coltrane. 2009. Divisions of labor in Mexican origin and Anglo families: Structure and culture. *Sex Roles* 60:482–95.

Schofield, T. J., R. D. Parke, Y. Kim, and S. Coltrane. 2008. Bridging the acculturation gap: Parent-child relationship quality as a moderator in Mexican American families. *Developmental Psychology* 44:1190–94.

"How Can You Do That to Your Children?"

Meredith Maran

One day when my sons Jesse and Peter were four and five years old, I looked around at our safe, all-white suburb of San Jose, California; at the landscaped, picture-perfect, nearly all-white elementary school my sons would soon be attending; at their nearly all-white friends—and I saw history repeating itself.

Forty years ago, I grew up in an all-white neighborhood in Manhattan. I went to an all-white elementary school. The first black person I ever saw was my second-grade teacher. I was twelve before I had a conversation with an African American my age. When Harlem erupted in the wake of Martin Luther King Jr.'s assassination, my father paced the floors at night, convinced that the "Negroes" were about to come marching downtown to get us.

I didn't want my sons to grow up in that state of isolation and ignorance. I didn't want to plant in my children the seeds of fear—the fear of the unknown that would bloom, later, into bigotry. And I didn't want to leave the world the way I'd found it by raising another generation in segregation. So in 1984 I moved to a mixed, mostly black Oakland neighborhood and enrolled my sons in a mixed, mostly black Oakland public school.

Years later, our house was often filled with teenage boys. Jesse's friends were all African American, all basketball players; they had Nike swooshes carved into their shaved heads and basketballs in their book bags. Peter's friends were Chilean and Mexican and African American; they had dreadlocks and wore Bob Marley T-shirts. On

Sunday mornings the boys all went rummaging through our fridge together—Jesse's friends looking for milk to pour on their Captain Crunch; Peter's friends looking for soy milk for their granola—and my heart would swell with joy. *This* is what I had in mind. *This* is how the world should be.

But sometimes in the midst of one of those warm moments, I was chilled by thoughts of the disparities between black boys and white boys. Today, African American students, mostly boys, are suspended and expelled at three times the rate of white ones. Forty-seven percent of black boys graduate from high school, compared to 75 percent of white male students. And in California today, 40 percent of African American males between the ages of eighteen and twenty-five are either in jail, on parole, or on probation.

I'd look around my living room at these young men I've loved since they were five or ten or fifteen years old, these boys who live in my heart as surely as if they'd once lived in my body, and I'd wonder: *which four won't make it?*

When my boys were in high school, I went on tour for my first book, *What It's Like to Live Now*, reading stories about my kids and our neighborhood. In every city, at nearly every reading, a white woman in the audience would raise her hand and ask the same question.

"How can you do that to your children?"

"Do what?" I'd ask, knowing what the answer would be: "raise them in the ghetto"; "send them to inner-city public schools"; "expose them to so much violence." None of the women said what she really meant: "subject your middle-class white kids to the dangers and deprivations that only poor black kids are supposed to be subjected to."

"Sometimes I question my decision," I'd answer honestly, thinking of the long nights when I sit by the phone with my heart in my throat because my kids are ten minutes late; and the anxious days when I look at the Cs on their report cards and wonder if private schools or suburban schools could have turned my boys into "college material."

And then I'd think about my sons' passion for justice; their intolerance of intolerance; their deep friendships with people of races and cultures different from their own, and how those friendships have rooted them in what they call (and shifting demographics indicate what is, in fact) "the real world."

"Mostly I'm glad I raised my kids the way I did," I'd say then. "Every parent has regrets. If I'd raised my kids in a white suburb, I'd have regrets about something else.

"I don't want my kids to grow up the way I did, in an all-white bubble. I think the best way to start solving the problems *every* parent worries about—crime, drugs, poverty, underfunded schools—is to start dealing with racism in society, and in ourselves. And the best way for white parents to do that is to start making different choices about how we raise our kids. The default mode is segregation. That's how we got where we are now."

This is what I said, and this is what I still believe. But I said it gently, with compassion, because I knew how difficult it is to be a white parent and *choose* to make the joys and suffering of African American children the joys and suffering of our own.

I know how wrenching it is to go to back-to-school night and listen to teachers begging for scrap paper and volunteers, instead of talking about grades and curriculum. I know how enriching *and* how painful it is to love black boys, to welcome them into my family and my heart, and watch them grow up with shopkeepers following them around in stores, cops shining flashlights in their faces, white people crossing the street to avoid them. I know the despair it evokes to be close enough to black boys to overhear the message they get on a daily basis: that nothing (or worse) is expected of them, and nothing (or worse) will be offered to them.

I know, too, that it benefits no one for white parents to "care about" the fate of black boys in the abstract, or moralistically, or patronizingly. What moved me to go to school board meetings and argue for smaller, more relevant classes, more guidance counselors, more nutritious school lunches; what moved me to stop and make my presence known when I see shopkeepers or police officers confronting African American boys; what moved me to work with my neighbors to put the crack dealer on our block out of business is that I had invested my children in the outcome.

Twenty-five years after I made the decision to cast my sons' fate with kids from a different race and class, the results are in. My boys are alive and doing well. Peter is a Los Angeles– and New York–based photographer shooting album packages and promo portraits for hip-hop artists. His Oakland upbringing gives him the street smarts he needs to relate to his subjects; his middle-class background gives him the "office smarts" he needs to relate to the executives and art directors who give him his jobs.

Jesse, so far, has stayed closer to home, metaphorically and geographically. He's a gifted visual artist and a dedicated social worker whose clients include drug-addicted teenagers, disenfranchised seniors, and homeless veterans. He's happily engaged to an African American woman who is the joy of his life, and our family's. Grandchildren? I'm waiting, optimistically.

But what about the boys they grew up with? Many of them, too many, are indeed dead, in jail, or on the streets. Because they're not African American, Jesse and Peter were grazed, not wounded, by the stray bullets of racism that were aimed at their friends. But my sons grew up in a mostly black neighborhood, and they were educated in mostly black public schools. So they are motivated, as I am, by love and by self-interest—a powerful combination—to raise the level of the river beneath the boat we're all in together.

Wherever and however we live, every child, every parent, every family can help bridge the racial gulf that divides our country, and divides us from each other. Parents who live in mostly white neighborhoods or send their kids to mostly white schools can do much to give their kids the values and the experience they'll need to live in an increasingly diverse world. Parents can teach tolerance at the dinner table, and buy their kids toys and take their kids to events that celebrate diversity. They can agitate at their kids' schools for multicultural school boards, faculty, curriculum, textbooks, library collections, and assemblies. They can urge their PTAs to cosponsor events with inner-city PTAs. They can get their kids on to racially mixed sports teams, and take their kids to racially mixed after-school care centers and religious services.

Recognizing these possibilities, as white parents we also recognize the responsibility we share: to stop being part of the problem and start being part of the solution. I have no doubt that our children are smart enough, resilient enough, and open-hearted enough to meet the challenge.

The question is, are we?

Me and My Nose

Rona Fernandez

When I was a child, no more than five years old, my mother would sing a peculiar lullaby to help put me to sleep. We would lie together on my bed, my head cradled on her arm, her breath warm and comforting on my forehead. My eyes would be closed, the room dark. I cherished these quiet moments of togetherness with my mom—a Filipina immigrant and single mother who worked so much that I only saw her at night after she picked me up from the babysitter. As I lay there, half-dozing beside her, she would gently pull on the bridge of my flat, Filipina nose, and sing, "Make your nose higher-higher, make your nose higher-higher."

For years afterward, I never thought there was anything strange or wrong with my mother's wish that my nose would miraculously sharpen and grow to be more like a European nose. Later, I would see how the adults in my family fawned over mestizo (mixed, usually half-white) children, remarking on how fair they were, and how pretty their lighter brown hair and pointy noses. I would also look at the bottles of Eskinol—a Filipino skin-whitening tonic—on my aunt's vanity table, the brunette model on the bottle's label as pale as the blond Barbie doll I played with every day, and recognize the model's ski-slope nose as the one that my mother had coveted for me.

Then, in 2007 I made my first trip to the Philippines, a mostly monoracial country where gradations of skin color and body weight are the key markers of one's social status. There I met two of my cousins for the first time, a young brother and sister who were beautiful to my Filipino American eyes, with their dark brown skin, slender physiques, and handsome features. They were as dark as many black people I knew back in the States. When I said they were pretty enough to be models, the rest of my family laughed.

"But they're so dark!" exclaimed a lighter-skinned cousin, incredulous that I could see beauty where they clearly saw none. To them, dark skin denoted a lack of education that forced people to earn their living working under the hot tropical sun. And slender meant you were skinny because you were poor and didn't have enough food to eat.

On the other hand, my Filipino family greeted me with delight. "You're so pretty and white! You looked so black in the pictures," they said, referring to photos they'd seen of me before my arrival. I recoiled at their words, having thought of myself as a person of color for so long that being labeled "white" felt like being seen as a space alien.

While few, if any, of my family in the Philippines have spent much time with non-Asians, I was born and raised in the Bay Area and call Oakland, California—one of the most diverse cities in the country—my home. As the American-born daughter of Filipino immigrant parents, I've struggled for years to overcome the various forms of racism at play in my life—not just those directed at me, but those I find within myself. I've found that overcoming my own internalized racism has been my first step toward a life less constrained by race. For many people of color, this type of racism is subtle but powerful, affecting our self-esteem in ways we are often not even aware. And it's not just about accepting ourselves—it's also about accepting our differences with other people, including other people of color.

For even when there are no white folks around, racism is a fact of life. I've learned this not just from my Filipino family members, but from my own circle of progressive friends. Like when I was in Chinatown with an African American friend, who looked up at some of the red and gold Chinese New Year decorations and asked me, as if I had some secret Asian knowledge, "Why do Chinese people like such ugly colors?"

I know I'm not immune to knee-jerk prejudiced reactions: one night, as my Chinese American husband and I arrived home, I noticed a group of three young black men playing dice in a small alcove in front of our neighbor's garage. I'd never seen them there before, didn't recognize them. My heart began beating faster, my throat tightened, and I started to wonder whether we were about to get robbed.

Nothing happened. But I still felt unsettled. Just a few days earlier, I'd heard about a neighbor who'd gotten robbed at gunpoint, and a friend who lived down the street who had his car stolen recently. Of course, I didn't know whether either of those incidents involved young black men, but in my head the connection was instantaneous, reflexive. And having many colleagues, neighbors, and close friends who are black hadn't done much to prevent these fears from coming up.

Although I knew I could be making myself look foolish and racist, I shared my fears with friends, all people of color, including a few African Americans, and tried to make sense of them. Should I be worried? Was I just being paranoid? How could I be street-smart without being downright racist? One of my friends, who grew up a few blocks away from where I live, offered the best perspective. A much taller woman than me, in her youth she would dress in baggy jeans and sweatshirts, the unofficial uniform of many young men of color in Oakland.

"When I would walk around by myself at night, dressed the way I was, and come across a young black man walking the other way, I could sense tension," she told me. "He would tighten up, because he thought I was a dude. As soon as he realized that I was a woman, his whole body would relax. He had been just as scared that I might do something to him as I might be scared of him doing something to me."

Her story reminded me that the supposedly "violent black men" that many of us fear have their own vulnerabilities, that statistically they are more at-risk of having a violent act perpetrated on them than I am. It reminded me of how ingrained those fears of violent young black men are in everyone's consciousness—even in the minds of young black men themselves, not to mention my own. I still have a lot of work to do to change these often unconscious beliefs and attitudes, despite the many years I've spent actively trying to unlearn racism.

It's a theme that pops up in psychological research about racism, in which our conscious minds seem to do constant battle with subconscious prejudice. A "slipup" is what psychologist David Amodio calls these automatic racial prejudices in his contribution to this anthology, "The Egalitarian Brain." Yet Amodio suggests that these slips can ultimately serve to combat prejudice, for they make us put more effort into changing future behavior. If we hold antiracist values, an occasional jolt of prejudiced impulse may just motivate us to examine our biases more closely, the better to guard against them.

This has been true for me. What were initially negative experiences of shame around my unresolved racial prejudice have become constructive ways to help me confront and move past these prejudices. The changes were subtle, to be sure—such as reminding myself of my friend's story when I see an unfamiliar young black man walking on my block, helping to ease the kind of anxiety I'd experienced that night in front of my house. But these changes have been more powerful and longer lasting than any intellectual shifts I've experienced after reading books on race or attending diversity trainings. I feel fortunate to live in a place where I interact with people of different racial backgrounds all the time, where I can learn from others' cultural traditions as well as teach them about mine. I've chatted on the bus with a Chinese immigrant woman about why I should learn Mandarin, and with a white woman who moved to Oakland from Savannah, Georgia, because she couldn't deal with Southern racism anymore. I've seen people of all colors getting along in Oakland more than I've seen the violence with which my adopted city is supposedly rife.

Within my family, race remains complex territory; I see signs of progress as well as stasis. At our wedding my husband and I chose our closest friends to stand with us as our wedding party. While three of our attendants were Filipina, five others were of African descent and one was Latino. Our officiant was a Chinese American Buddhist woman, a real shock to my Catholic relatives. Afterward, people made coded remarks like, "There weren't a lot of Filipinos at your wedding" —meaning, "Why were there so many black people?" My mother, bless her, was more positive and upbeat about it, proclaiming loudly to anyone who would listen that our wedding had been so "different" and "international." (Of course, very few of my friends who were people of color were born outside the United States.)

Perhaps my mother has evolved; I'd like to think so.

Then I think about the children I know who are growing up in a world where we not only have a black president but also people of different races and ethnicities starring in a range of top TV shows and movies. I think about how my friend's African American toddler asks for *agua* instead of water, or how I crave jalapeño chilies when I travel outside of California and our Mexican-influenced culinary culture. I think of the club I went to a few weeks ago where a predominantly black and Asian crowd grooved to house music in a downtown Oakland loft. I realize how different this world is than even the world I grew up in.

This is what research discussed by Amodio (and others in this anthology) predicts should happen: As our exposure to racial differences increases—and if we make a conscious commitment to acknowledge our automatic prejudices and to overcome them—we can train ourselves to be multicultural beings, or, to put it a different way, to learn to embrace rather than fear differences, the ones we find inside and the ones we find outside. My nose isn't any higher today than when I was a little girl, but now I think it's beautiful. That's a start.

Double Blood

Rebecca Walker

Like many biracial Americans of my generation, my parents met in the tumultuous cultural revolution of the 1960s. They married when it was illegal for people of different races to do so, and continued to challenge entrenched assumptions about race by having me. It was dangerous work. The Klan threatened our interracial family in Mississippi often. My father's Jewish mother disowned him for marrying a black woman.

Thirty-something years later, my copper-colored self has navigated bar mitzvahs, barbecues, inner-city public schools, the Ivy League, and a host of other wildly divergent worlds. I've been romantic and platonic with mono- and multiracial people in half a dozen American cities. And because of the memoir I wrote in my late twenties about being biracial, I've spoken intimately with multiracial individuals all over the world.

The pliability of racial and cultural identity can be a tremendous gift. In Kenya, Mexico, Thailand, Egypt, Morocco, and several countries in between, I am embraced as a daughter, sister, mother, or potential wife. Rarely, if ever, am I perceived as a foreigner. "Ah," the shopkeeper, teacher, or taxi driver will say, "you come from America? But you are like us!" I laugh and shake my head. "No matter," the person will say, "now you are home." And I will feel at home, too. Even if I cannot speak the language, I easily absorb the mannerisms and rhythm of daily life.

Multiracial friends share similar experiences. Within hours, even minutes, of being immersed in another culture, our "own" ethnicity seems to disappear. A Swedish-Nigerian woman I spoke to recently told me her biracial identity made her a border crosser and, as she put it, a "walking heart."

The not-so-fabulous part of being multiracial is the tremendous anxiety often felt as a result of belonging to several racial and cultural groups, each demanding almost lockstep allegiance. Should hair be straightened or twisted into dreadlocks? What slang is to be used, literary legacy privileged, political candidate supported? A decision as simple as which neighborhood to live in can be a minefield for a multiracial person pressured to embrace one race or culture.

In recent years researchers have started to focus on the multiracial experience in all its complexity. Their work has documented the particular breeds of prejudice and anxiety we can face. But it also offers hope that as the next generation of multiracial families makes its way in the world, they're developing new strategies to meet these challenges—strategies that the pioneers of previous generations never quite figured out.

Perhaps because I know the biracial predicament intimately, much of this research rings especially true for me. In one study sociologist Kimberly Brackett and her colleagues found that multiracial people experience more racism than our monoracial counterparts. In another, Yoonsun Choi at the University of Chicago found multiracial youth at greater risk for substance abuse and violent behavior than monoracial youth of the same socioeconomic backgrounds.

These findings reflect my own experience precisely. Always too black or too white, I was privy to the prejudices of both racial groups, and I was used as a screen onto which each projected their idea of the Other. Though I had to countermand white racism, I endured slights from African Americans as well. My mannerisms were labeled snobby and standoffish. Even today within African American academic and intellectual circles my credentials are often in question. The "privilege" of being half-white is perceived as a kind of "pass," a disqualifier of true intellectual merit.

When I was a teenager, I used drugs to bridge the gaps between communities and to numb the pain of their judgment. Marijuana and ecstasy were great equalizers, a common language allowing for a temporary suspension of superficial differences. As a woman, I also found solace through sexual activity in which my skin color meant less than my viability as a sexual partner.

I know that many biracial and multiracial men use physical aggression and other forms of self-destructive behavior to prove their worth, a pattern that transcends racial categories. Several of the multiracial men I knew in high school, especially those from lower socioeconomic backgrounds, are now in prison. Another is dead.

But there is promising news from the multiracial front. In the last few years, biracial and multiracial people have begun to tell me in no uncertain terms about their growing ease with their dual or triple heritages. Though the number of Americans of "two or more races" is still below 5 million, according to the U.S. Census Bureau's Population Estimate Program, this population grew by 25 percent from 2000 to 2007; the total U.S. population grew by just 7 percent over the same period. As the mixed-race population expands—and certainly now that we have a biracial president—feelings of alienation and invisibility seem to be declining, often replaced by unabashed pride. I've been especially taken with the number of multiracial groups on college campuses. I struggled mightily to forge cross-cultural allegiances at Yale during my undergraduate years. Fifteen years later I find cohesive student groups of biracial and multiracial people on every campus I am fortunate enough to visit.

In addition to external supports, it appears parents are slowly gaining the facility to discuss these issues at home. A study by psychologist Barbra Fletcher Stephens has found that biracial couples are starting to acknowledge the importance of talking to their children about their biracial identity, and of "passing on the positive as well as the negative attitudes of both sides of their family in support of helping their children to develop a cohesive personal identity." In other words, it is important for parents to disclose both the pretty and not-so-pretty aspects of their cultural histories, rather than censoring or minimizing less-appealing aspects. Fletcher Stephens's study concludes that biracial and multiracial children with parents who present this kind of complex view of their family culture have less core identity conflict, and are thus better equipped to manage social environments still mired in prejudice and divisiveness.

As pioneers, the parents of my biracial and multiracial peers had few real studies, and even less popular discourse, on the subject from which to draw. As a result, they often discounted, minimized, or were completely unaware of their children's struggles with identity and prejudice. This new research brings good news—and proven problem-solving techniques—to this next generation of parents, who then have a better chance of identifying and avoiding the potential pitfalls awaiting their children. Growing familiarity with the topic also provides validation and visibility to biracial and multiracial children who may not yet be ready to speak about the prejudice they face in the classroom and on the playground. This is exceptionally good news for these children.

A detailed study of biracial families nationwide, published in a 2007 issue of the *American Journal of Sociology*, offers further evidence of the progress multiracial families are making, while also pointing to the challenges they still confront. Sociologists Simon Cheng and Brian Powell determined that parents in biracial families usually allocate more financial and cultural resources, such as music lessons and museum trips, to their children's education than do parents in corresponding monoracial families. However, the researchers also conclude that deeply entrenched social prejudices against interracial marriage, especially marriages between black men and white women, may leave these families socially isolated. Cheng and Powell surmise that, though these biracial families do not necessarily have more money, they might invest more resources in their children's education in an effort to compensate for their marginalized social position.

This study highlights a tension—between the evolving attitudes of multiracial families and individuals, and the prejudices that still persist in society at large—that I also recognized in an older study, one that struck me as particularly relevant for the future of biracial identity. In the early 90s, Stanford University psychologist Teresa LaFramboise and colleagues explored several approaches people take to resolve the psychological dissonance of their multicultural experience. These approaches range from assimilation, where people try to blend in completely with the culture they see as most desirable, to fusion, where they try to merge different cultures to form a new one. But the researchers saw special promise in what's called the "alternation" model, in which a biracial or bicultural person is able to fully function within two cultures without losing his or her identity, or having to choose one culture over the other.

I can see the allure of this model—the biracial person as masterful toggler, able to switch back and forth at the speed of light and remain unscathed by the transition. It's a model I employ successfully at times and find inspiring to observe in others. I spoke recently at an arts festival in Amsterdam, for example, and found myself in deep discussion with my Dutch-Surinamese counterparts. When I stumbled over appropriate terminology, they emphatically shared what they've come to. "*Double Blood*," they told me. "That's what we are. Neither one nor the other, neither better nor worse. We are both, we are proud, and we are beautiful." I was literally almost knocked over by the sheer force of their self-certainty, and adopted the term immediately.

I am still concerned, though, that even this most reasonable of solutions places the burden of psychological transformation on biracial people themselves rather than on the society at large. Biracial and multiracial people are expected to find a way to mediate a world seen as inherently and irreversibly divided. They must learn to walk skillfully between worlds deemed multiple, rather than peaceably through a world they experience as one.

This may serve as a useful coping strategy, but it isn't a sustainable long-term plan for ending prejudice in our society. That will come only once all human beings reckon with the good and bad of their ancestral legacy and work to transform the rhetoric of separate and unequal into one of united and open to all.

Luckily, through this new research on biracial and multiracial families, we are that much closer to a road map for making this cultural shift. It will require that all parents talk to their children about their cultural legacy—both positive and appalling. Only once they're so informed can children begin to reconcile age-old racial conflicts, including conflicts in their own identity.

Moving in this way, we teach our children that the past is known, but the future is free, unwritten, and dependent upon them. This idea alone can shape a generation, and make prejudice a thing of the past.

Truth + Reconciliation

Archbishop Desmond Tutu

Malusi Mpumlwana was a young enthusiastic antiapartheid activist and a close associate of Steve Biko in South Africa's crucial Black Consciousness movement of the late 1970s and early 1980s. He was involved in vital community development and health projects with impoverished and often demoralized rural communities. As a result, he and his wife were under strict surveillance, constantly harassed by the ubiquitous security police. They were frequently held in detention without trial.

I remember well a day Malusi gave the security police the slip and came to my office in Johannesburg, where I was serving as general secretary of the South African Council of Churches. He told me that during his frequent stints in detention, when the security police routinely tortured him, he used to think, "These are God's children and yet they are behaving like animals. They need us to help them recover the humanity they have lost." For our struggle against apartheid to be successful, it required remarkable young people like Malusi.

All South Africans were less than whole because of apartheid. Blacks suffered years of cruelty and oppression, while many privileged whites became more uncaring, less compassionate, less humane, and therefore less human. Yet during these years of suffering and inequality, each South African's humanity was still tied to that of all others, white or black, friend or enemy. For our own dignity can only be measured in the way we treat others. This was Malusi's extraordinary insight.

I saw the power of this idea when I was serving as chairman of the Truth and Reconciliation Commission in South Africa. This was the commission that the postapartheid government, headed by our president, Nelson Mandela, had established to move us beyond the cycles of retribution and violence that had plagued so many other countries during their transitions from oppression to democracy. The commission granted perpetrators of political crimes the opportunity to appeal for amnesty by giving a full and truthful account of their actions and, if they so chose, an opportunity to ask for forgiveness—opportunities that some took and others did not. The commission also gave victims of political crimes a chance to tell their stories, hear confessions, and thus unburden themselves from the pain and suffering they had experienced.

For our nation to heal and become a more humane place, we had to embrace our enemies as well as our friends. The same is true the world over. True enduring peace—between countries, within a country, within a community, within a family—requires real reconciliation between former enemies and even between loved ones who have struggled with one another.

How could anyone really think that true reconciliation could avoid a proper confrontation? After a husband and wife or two friends have quarreled, if they merely seek to gloss over their differences or metaphorically paper over the cracks, they must not be surprised when they are soon at it again, perhaps more violently than before, because they have tried to heal their ailment lightly.

True reconciliation is based on forgiveness, and forgiveness is based on true confession, and confession is based on penitence, on contrition, on sorrow for what you have done. We know that when a husband and wife have quarreled, one of them must be ready to say the most difficult words in any language, "I'm sorry," and the other must be ready to forgive for there to be a future for their relationship. This is true between parents and children, between siblings, between neighbors, and between friends. Equally, confession, forgiveness, and reconciliation in the lives of nations are not just airy-fairy religious and spiritual things, nebulous and unrealistic. They are the stuff of practical politics.

Those who forget the past, as many have pointed out, are doomed to repeat it. Just in terms of human psychology, we in South Africa knew that to have blanket amnesty where no disclosure was made would not deal with our past. It is not dealing with the past to say glibly, "Let bygones be bygones," for then they will never be bygones. How can you forgive if you do not know what or whom to forgive? In our commission hearings, we required full disclosure for us to grant amnesty. Only then, we thought, would the process of requesting and receiving forgiveness be healing and transformative for all involved. The commission's record shows that its standards for disclosure and amnesty were high indeed: of the more than 7,000 applications submitted to the Truth and Reconciliation Commission, it granted amnesty to only 849 of them.

Unearthing the truth was necessary not only for the victims to heal, but for the perpetrators as well. Guilt, even unacknowledged guilt, has a negative effect on the guilty. One day it will come out in some form or another. We must be radical. We must go to the root, remove that which is festering, cleanse and cauterize, and then a new beginning is possible.

Forgiveness gives us the capacity to make a new start. That is the power, the rationale, of confession and forgiveness. It is to say, "I have fallen but I am not going to remain there. Please forgive me." And forgiveness is the grace by which you enable the other person to get up, and get up with dignity, to begin anew. Not to forgive leads to bitterness and hatred, which just like self-hatred and self-contempt, gnaw away at the vitals of one's being. Whether hatred is projected out or projected in, it is always corrosive of the human spirit.

We have all experienced how much better we feel after apologies are made and accepted, but even still it is so hard for us to say that we are sorry. I often find it difficult to say these words to my wife in the intimacy and love of our bedroom. How much more difficult it is to say these words to our friends, our neighbors, and our coworkers. Asking for forgiveness requires that we take responsibility for our part in the rupture that has occurred in the relationship. We can always make excuses for ourselves and find justifications for our actions, however contorted, but we know that these keep us locked in the prison of blame and shame.

In the story of Adam and Eve, the Bible reminds us of how easy it is to blame others. When God confronted Adam about eating the forbidden fruit from the Tree of Knowledge of Good and Evil, Adam was less than forthcoming in accepting responsibility. Instead he shifted the blame to Eve, and when God turned to Eve, she too tried to pass the buck to the serpent. (The poor serpent had no one left to blame.) So we should not be surprised at how reluctant most people are to acknowledge their responsibility and to say they are sorry. We are behaving true to our ancestors when we blame everyone and everything except ourselves. It is the everyday heroic act that says, "It's my fault. I'm sorry." But without these simple words, forgiveness is much more difficult.

Forgiving and being reconciled to our enemies or our loved ones are not about pretending that things are other than they are. It is not about patting one another on the back and turning a blind eye to the wrong. True reconciliation exposes the awfulness, the abuse, the pain, the hurt, the truth. It could even sometimes make things worse. It is a risky undertaking but in the end it is worthwhile, because in the end only an honest confrontation with reality can bring real healing. Superficial reconciliation can bring only superficial healing.

If the wrongdoer has come to the point of realizing his wrong, then one hopes there will be contrition, or at least some remorse or sorrow. This should lead him to confess the wrong he has done and ask for forgiveness. It obviously requires a fair measure of humility. But what happens when such contrition or confession is lacking? Must the victim be dependent on these before she can forgive? There is no question that such a confession is a very great help to the one who wants to forgive, but it is not absolutely indispensable. If the victim could forgive only when the culprit confessed, then the victim would be locked into the culprit's whim, locked into victimhood, no matter her own attitude or intention. That would be palpably unjust.

In the act of forgiveness, we are declaring our faith in the future of a relationship and in the capacity of the wrongdoer to change. We are welcoming a chance to make a new beginning. Because we are not infallible, because we will hurt especially the ones we love by some wrong, we will always need a process of forgiveness and reconciliation to deal with those unfortunate yet all too human breaches in relationships. They are an inescapable characteristic of the human condition.

We have had a jurisprudence, a penology in Africa that was not retributive but restorative. Traditionally, when people quarreled the main intention was not to punish the miscreant but to restore good relations. This was the animating principle of our Truth and Reconciliation Commission. For Africa is concerned, or has traditionally been concerned, about the wholeness of relationships. That is something we need in this world—a world that is polarized, a world that is fragmented, a world that destroys people. It is also something we need in our families and friendships. For retribution wounds and divides us from one another. Only restoration can heal us and make us whole. And only forgiveness enables us to restore trust and compassion to our relationships. If peace is our goal, there can be no future without forgiveness.

Acknowledgments

Most of the contributions to this book originally appeared in *Greater Good* magazine (www.greatergoodmag.org), which has flourished for more than six years because of the hard work of hundreds of people, from staff and volunteers to writers and editorial board members. We extend our most profound gratitude to all of our dedicated readers and donors, even if space limitations prevent us from thanking each by name.

We would like to express special gratitude to Thomas and Ruth Ann Hornaday and the Herb Alpert Foundation, whose combined support made *Greater Good* possible, and to Lee Hwang and his Quality of Life Foundation, whose support has helped the magazine expand its reach.

We also owe huge thanks to Dacher Keltner, *Greater Good*'s executive editor and the founding faculty director of the Greater Good Science Center, for his vision and leadership, which brought *Greater Good* into being.

We are deeply indebted to the staff at Beacon Press, including its director, Helene Atwan. Andy Hrycyna at Beacon served as an early champion of *Are We Born Racist?* Alex Kapitan improved the volume at every stage of its development with her thoughtful edits.

This book's development also benefited immensely from input from leading scientific experts on racism and prejudice. We're especially thankful to Elizabeth Page-Gould, Jack Glaser, Susan Fiske, Barbara Fredrickson, Samuel Gaertner, Walter Stephan, and Thomas Pettigrew for their time and assistance.

Finally, we are grateful to Meredith Milet, Ozlem Ayduk, and Olli Doo for their patience and support.

Resources

Antiracist community organizing

The Leadership Advocacy Program of the Applied Research Center strengthens the capacity of community-based organizations, training intermediaries, and individual activists to engage in education and action to advance racial equity. The Applied Research Center also publishes *ColorLines*, a national news magazine on race and politics. **www.arc.org**.

The Center for Community Change helps low-income people, especially people of color, build powerful organizations through which they can change their communities and public policies for the better. **www.communitychange.org**.

Antiracist curricula

Teaching Tolerance offers free resources—including Oscar-winning film kits—that can help educators effectively address social problems like racism. **www.tolerance.org**.

Facing History and Ourselves offers innovative, relevant materials that engage students in an examination of racism, prejudice, and anti-Semitism, as well as professional development opportunities for educators. **www.facinghistory.org**, 617-232-0281.

Teaching for Change is a leading distributor of antibias and antiracist resources and seeks to "transform schools into centers of justice where students learn to read, write and change the world." **www.teachingforchange.org**, 800–763–9131.

Character education

The Character Education Partnership focuses on defining and encouraging effective practices and approaches to quality character education. Its Web site offers free, downloadable resources. **www.character.org**, 800-988–8081.

Everyday Democracy (formerly the Study Circles Resource Center) offers "dialogue-to-change" programs that can help students examine racism in their schools and communities and develop solutions and implement them. **www.everyday-democracy.org**, 860–928–2616.

Collaborations between law enforcement and scientists

The Consortium for Police Leadership in Equity is a research consortium that promotes police transparency and accountability by facilitating innovative research collaborations between law enforcement agencies and world-class social scientists. **http://cple.psych.ucla.edu**.

The **Policing Racial Bias** project aims to develop partnerships among social psychologists and law enforcement agencies to share information and generate new knowledge on the influence of racial bias in policing. **http://waldron.stanford.edu/~policingproject**.

Cross-group contact and cooperative learning

Teaching Tolerance sponsors **Mix It Up at Lunch Day** each November, during which students take a new seat in the cafeteria and get to know someone new. Research shows the program helps students become more comfortable interacting with different kinds of people, bolsters school unity, and fosters new friendships. **www.mixitup.org**.

The Success for All Foundation helps schools and educators learn how to use cooperative learning strategies to support student achievement. **www.successforall.net**, 800–548–4998.

The Jigsaw Classroom is a cooperative learning technique with a three-decade track record of successfully reducing racial conflict and increasing positive educational outcomes. **www.jigsaw.org**.

Personal prejudice tests

Project Implicit offers a variety of online tests that measure conscious and unconscious preferences, including racial prejudices. Users can demo these tests or have their scores included in psychological studies. **https://implicit.harvard.edu/implicit**.

The shooting simulation used by University of Chicago psychologist Joshua Correll, in which users have 850 milliseconds to decide whether to shoot white or African American men holding guns or harmless objects, is online at **http://backhand.uchicago.edu/Center/ShooterEffect**.

Resources for parents and kids

Anti-Racist Parent is a blog and resource for parents who are committed to raising children with an antiracist outlook. **www.antiracistparent.com**.

Project RACE advocates for multiracial children and adults through education, community awareness, and legislation. Teen Project RACE gives multiracial youth a way to advocate for themselves. **www.projectrace.com**.

School equity

Run by current and former classroom teachers, **Rethinking Schools** works to advance equitable reform throughout the public school system in the United States. **www.rethinkingschools.org**, 414–964–9646.

The Center for Research on Education, Diversity and Excellence is a federally funded research and development program that works to improve education for students whose ability to reach their potential is challenged by language or cultural barriers, race, geographic location, or poverty. **http://crede.berkeley.edu**, 510–643–9024.

Contributors

David Amodio, PhD, is an assistant professor of psychology at New York University, where he researches the roles of social cognition and emotion in the regulation of behavior, and the neural mechanisms underlying these processes.

Dottie Blais teaches English and creative writing at Gainesville State College in Georgia. Previously she taught secondary school English in Georgia, Louisiana, and Colorado for more than twenty years.

Allison Briscoe-Smith, PhD, is a professor at Pacific Graduate School of Psychology. Both her research and clinical work focus on parenting and racial identity development.

Jennifer A. Chatman, PhD, is the Paul J. Cortese Distinguished Professor of Management at the Haas School of Business, University of California, Berkeley.

Scott Coltrane, PhD, is a professor of sociology at the University of Oregon, where he serves as dean of the college of arts and sciences. He is the author of *Family Man* (Oxford University Press, 1996) and, with Michele Adams Wadsworth, *Gender and Families* (Rowman & Littlefield, 2008), among other books.

Alex Dixon is an editorial assistant at *Greater Good*, the magazine of the Greater Good Science Center at the University of California, Berkeley.

Eve Ekman, MSW, is a San Francisco–based social worker, writer, artist, and editor of the online interdisciplinary magazine *Ethsix*: Social Work, Journalism, and Art from the Invisible City.*

Rona Fernandez, a second-generation Filipina American, is a fundraiser, activist, and writer based in Oakland, California.

Susan T. Fiske, PhD, is Eugene Higgins Professor of Psychology at Princeton University and the author, with Shelley E. Taylor, of *Social Cognition, from Brains to Culture* (McGraw-Hill, 2007), among other books.

Anita Foeman, PhD, is a professor of communication studies at West Chester University, where she specializes in intercultural and organizational communication.

Jennifer Holladay, MS, is the former director of the Teaching Tolerance project at the Southern Poverty Law Center.

Kareem Johnson, PhD, is an assistant professor of psychology at Temple University, where he studies the emotional, cognitive, and psychophysiological factors that may influence how people perceive and respond to outgroup members.

Meredith Maran is the author or coauthor of eight books, including *Dirty: A Search for Answers inside America's Teenage Drug Epidemic* (HarperOne, 2003) and *Class Dismissed: A Year in the Life of an American High School, A Glimpse into the Heart of a Nation* (St. Martin's Griffin, 2001). A different version of her contribution to this anthology was originally published in *Parenting* magazine.

Jason Marsh is the editor in chief of *Greater Good* magazine and coeditor, with Dacher Keltner and Jeremy Adam Smith, of *The Compassionate Instinct* (W. W. Norton, 2010). Previously he was the managing editor of the journal the *Responsive Community.* He is also the producer of the documentary film *Unschooled* and has worked as a public radio reporter and producer.

Rodolfo Mendoza-Denton, PhD, is an associate professor of psychology at the University of California, Berkeley, where he studies stereotyping and prejudice from the perspective of both target and perceiver, intergroup relations, and cultural psychology.

Terry Nance, PhD, is a professor of communication studies at Villanova University, where she serves as assistant vice president for multicultural affairs.

Elizabeth Page-Gould, PhD, is an assistant professor of psychology at the University of Toronto and a former postdoctoral fellow at the Mind/Brain/Behavior Initiative at Harvard University, where she studied cross-group friendship, intergroup relations, social interaction, social cognition, and psychophysiology.

Ross D. Parke, PhD, is a distinguished professor emeritus of psychology and the former director of the Center for Family Studies at the University of California, Riverside. He is the author of *Fatherhood* (Harvard University Press, 1996) and, with Armin Brott, *Throwaway Dads* (Houghton Mifflin, 1999), as well as other books on fatherhood and child psychology.

Thomas Schofield, PhD, is a postdoctoral fellow at the University of California, Davis, who studies family processes and child development in different cultural contexts. His work has appeared in journals (*Developmental Psychology* and *Journal of Nonverbal Behavior*) as well as edited volumes.

Jeremy Adam Smith is the author of *The Daddy Shift* (Beacon Press, 2009); coeditor, with Dacher Keltner and Jason Marsh, of *The Compassionate Instinct* (W.W. Norton, 2010); and editor of the online magazine Shareable.net. His articles and essays on parenting, urban life, race, politics, and technology have appeared in the *Nation, San Francisco Chronicle, San Francisco Bay Guardian, Utne Reader, Wired,* and numerous other periodicals and anthologies. He serves as contributing editor of *Greater Good* magazine, where he is also the former senior editor.

Desmond Tutu is the recipient of the 1984 Nobel Peace Prize. After he retired as archbishop of Cape Town, South Africa, in 1996, Tutu served as chairman of South Africa's Truth and Reconciliation Commission. His contribution to this anthology was drawn from his 2004 book, *God Has a Dream* (Doubleday), and was published in the fall 2004 issue of *Greater Good* magazine.

Rebecca Walker is an author, editor, speaker, and teacher. Her first memoir, *Black, White and Jewish: Autobiography of a Shifting Self* (Riverhead, 2001), is a national best seller and winner of the Alex Award from the American Library Association. She is also the author of *Baby Love: Choosing Motherhood after a Lifetime of Ambivalence* (Riverhead, 2007) and editor of the anthology *One Big Happy Family* (Riverhead, 2009).